Lecture Notes in Mathematics

Edited by A. Dold and B. Eckmann

1267

J. Lindenstrauss V. D. Milman (Eds.)

Geometrical Aspects of Functional Analysis

Israel Seminar, 1985–86

Springer-Verlag
Berlin Heidelberg New York London Paris Tokyo

Editors

Joram Lindenstrauss
Hebrew University of Jerusalem
Givat Ram, 91 904 Jerusalem, Israel

Vitali D. Milman
School of Mathematical Sciences, Tel Aviv University
Ramat Aviv, 69 978 Tel Aviv, Israel

Mathematics Subject Classification (1980): 46B20, 52A20

ISBN 3-540-18103-2 Springer-Verlag Berlin Heidelberg New York
ISBN 0-387-18103-2 Springer-Verlag New York Berlin Heidelberg

Printing and binding: Druckhaus Beltz, Hemsbach/Bergstr.
2146/3140-543210

Lecture Notes in Mathematics

Edited by A. Dold and B. Eckmann

1267

J. Lindenstrauss V. D. Milman (Eds.)

Geometrical Aspects of Functional Analysis

Israel Seminar, 1985–86

Springer-Verlag

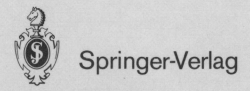

Lecture Notes in Mathematics

For information about Vols. 1–1062 please contact your bookseller or Springer-Verlag.

Vol. 1063: Orienting Polymers. Proceedings, 1983. Edited by J.L. Ericksen. VII, 166 pages. 1984.

Vol. 1064: Probability Measures on Groups VII. Proceedings, 1983. Edited by H. Heyer. X, 588 pages. 1984.

Vol. 1065: A. Cuyt, Padé Approximants for Operators: Theory and Applications. IX, 138 pages. 1984.

Vol. 1066: Numerical Analysis. Proceedings, 1983. Edited by D.F. Griffiths. XI, 275 pages. 1984.

Vol. 1067: Yasuo Okuyama, Absolute Summability of Fourier Series and Orthogonal Series. VI, 118 pages. 1984.

Vol. 1068: Number Theory, Noordwijkerhout 1983. Proceedings. Edited by H. Jager. V, 296 pages. 1984.

Vol. 1069: M. Kreck, Bordism of Diffeomorphisms and Related Topics. III, 144 pages. 1984.

Vol. 1070: Interpolation Spaces and Allied Topics in Analysis. Proceedings, 1983. Edited by M. Cwikel and J. Peetre. III, 239 pages. 1984.

Vol. 1071: Padé Approximation and its Applications, Bad Honnef 1983. Prodeedings. Edited by H. Werner and H.J. Bünger. VI, 264 pages. 1984.

Vol. 1072: F. Rothe, Global Solutions of Reaction-Diffusion Systems. V, 216 pages. 1984.

Vol. 1073: Graph Theory, Singapore 1983. Proceedings. Edited by K.M. Koh and H.P. Yap. XIII, 335 pages. 1984.

Vol. 1074: E.W. Stredulinsky, Weighted Inequalities and Degenerate Elliptic Partial Differential Equations. III, 143 pages. 1984.

Vol. 1075: H. Majima, Asymptotic Analysis for Integrable Connections with Irregular Singular Points. IX, 159 pages. 1984.

Vol. 1076: Infinite-Dimensional Systems. Proceedings, 1983. Edited by F. Kappel and W. Schappacher. VII, 278 pages. 1984.

Vol. 1077: Lie Group Representations III. Proceedings, 1982–1983. Edited by R. Herb, R. Johnson, R. Lipsman, J. Rosenberg. XI, 454 pages. 1984.

Vol. 1078: A.J.E.M. Janssen, P. van der Steen, Integration Theory. V, 224 pages. 1984.

Vol. 1079: W. Ruppert. Compact Semitopological Semigroups: An Intrinsic Theory. V, 260 pages. 1984

Vol. 1080: Probability Theory on Vector Spaces III. Proceedings, 1983. Edited by D. Szynal and A. Weron. V, 373 pages. 1984.

Vol. 1081: D. Benson, Modular Representation Theory: New Trends and Methods. XI, 231 pages. 1984.

Vol. 1082: C.-G. Schmidt, Arithmetik Abelscher Varietäten mit komplexer Multiplikation. X, 96 Seiten. 1984.

Vol. 1083: D. Bump, Automorphic Forms on GL (3,IR). XI, 184 pages. 1984.

Vol. 1084: D. Kletzing, Structure and Representations of Q-Groups. VI, 290 pages. 1984.

Vol. 1085: G.K. Immink, Asymptotics of Analytic Difference Equations. V, 134 pages. 1984.

Vol. 1086: Sensitivity of Functionals with Applications to Engineering Sciences. Proceedings, 1983. Edited by V. Komkov. V, 130 pages. 1984

Vol. 1087: W. Narkiewicz, Uniform Distribution of Sequences of Integers in Residue Classes. VIII, 125 pages. 1984.

Vol. 1088: A.V. Kakosyan, L.B. Klebanov, J.A. Melamed, Characterization of Distributions by the Method of Intensively Monotone Operators. X, 175 pages. 1984.

Vol. 1089: Measure Theory, Oberwolfach 1983. Proceedings. Edited by D. Kölzow and D. Maharam-Stone. XIII, 327 pages. 1984.

Vol. 1090: Differential Geometry of Submanifolds. Proceedings, 1984. Edited by K. Kenmotsu. VI, 132 pages. 1984.

Vol. 1091: Multifunctions and Integrands. Proceedings, 1983. Edited by G. Salinetti. V, 234 pages. 1984.

Vol. 1092: Complete Intersections. Seminar, 1983. Edited by S. Greco and R. Strano. VII, 299 pages. 1984.

Vol. 1093: A. Prestel, Lectures on Formally Real Fields. XI, 125 pages. 1984.

Vol. 1094: Analyse Complexe. Proceedings, 1983. Edité par E. Amar, R. Gay et Nguyen Thanh Van. IX, 184 pages. 1984.

Vol. 1095: Stochastic Analysis and Applications. Proceedings, 1983. Edited by A. Truman and D. Williams. V, 199 pages. 1984.

Vol. 1096: Théorie du Potentiel. Proceedings, 1983. Edité par G. Mokobodzki et D. Pinchon. IX, 601 pages. 1984.

Vol. 1097: R.M. Dudley, H. Kunita, F. Ledrappier, École d'Été de Probabilités de Saint-Flour XII – 1982. Edité par P.L. Hennequin. X, 396 pages. 1984.

Vol. 1098: Groups – Korea 1983. Proceedings. Edited by A. C. Kim and B.H. Neumann. VII, 183 pages. 1984.

Vol. 1099: C.M. Ringel, Tame Algebras and Integral Quadratic Forms. XIII, 376 pages. 1984.

Vol. 1100: V. Ivrii, Precise Spectral Asymptotics for Elliptic Operators Acting in Fiberings over Manifolds with Boundary. V, 237 pages. 1984.

Vol. 1101: V. Cossart, J. Giraud, U. Orbanz, Resolution of Surface Singularities. Seminar. VII, 132 pages. 1984.

Vol. 1102: A. Verona, Stratified Mappings – Structure and Triangulability. IX, 160 pages. 1984.

Vol. 1103: Models and Sets. Proceedings, Logic Colloquium, 1983, Part I. Edited by G.H. Müller and M.M. Richter. VIII, 484 pages. 1984.

Vol. 1104: Computation and Proof Theory. Proceedings, Logic Colloquium, 1983, Part II. Edited by M.M. Richter, E. Börger, W. Oberschelp, B. Schinzel and W. Thomas. VIII, 475 pages. 1984.

Vol. 1105: Rational Approximation and Interpolation. Proceedings, 1983. Edited by P.R. Graves-Morris, E.B. Saff and R.S. Varga. XII, 528 pages. 1984.

Vol. 1106: C.T. Chong, Techniques of Admissible Recursion Theory. IX, 214 pages. 1984.

Vol. 1107: Nonlinear Analysis and Optimization. Proceedings, 1982. Edited by C. Vinti. V, 224 pages. 1984.

Vol. 1108: Global Analysis – Studies and Applications I. Edited by Yu.G. Borisovich and Yu.E. Gliklikh. V, 301 pages. 1984.

Vol. 1109: Stochastic Aspects of Classical and Quantum Systems. Proceedings, 1983. Edited by S. Albeverio, P. Combe and M. Sirugue-Collin. IX, 227 pages. 1985.

Vol. 1110: R. Jajte, Strong Limit Theorems in Non-Commutative Probability. VI, 152 pages. 1985.

Vol. 1111: Arbeitstagung Bonn 1984. Proceedings. Edited by F. Hirzebruch, J. Schwermer and S. Suter. V, 481 pages. 1985.

Vol. 1112: Products of Conjugacy Classes in Groups. Edited by Z. Arad and M. Herzog. V, 244 pages. 1985.

Vol. 1113: P. Antosik, C. Swartz, Matrix Methods in Analysis. IV, 114 pages. 1985.

Vol. 1114: Zahlentheoretische Analysis. Seminar. Herausgegeben von E. Hlawka. V, 157 Seiten. 1985.

Vol. 1115: J. Moulin Ollagnier, Ergodic Theory and Statistical Mechanics. VI, 147 pages. 1985.

Vol. 1116: S. Stolz, Hochzusammenhängende Mannigfaltigkeiten und ihre Ränder. XXIII, 134 Seiten. 1985.

continued on page 215

FOREWORD

These are the proceedings of the Israel Seminar on the Geometric Aspects of Functional Analysis which was held between October 1985 and June 1986. The seminar, which is supported by the Israel Mathematical Union has been held since 1980. The first session for which proceedings were published was the 1983-84 session. That year the proceedings were published privately by Tel Aviv University. The present volume is the second published proceedings of the seminar.

As the name indicates, the subject of the seminar is geometric problems in functional analysis in general and Banach space theory in particular. The study of convex sets in $\mathbb{R}^n$ and infinite-dimensional spaces is naturally a central topic in the subject. The wide scope of this research direction is, hopefully, clearly reflected in the papers of this volume.

The large majority of the papers in this volume are original research papers, reporting on recent results which have not been published elsewhere in the literature. The other papers are of an expository nature.

The list of the talks which were actually delivered at the seminar is given below. Not all of the talks given at the seminar are published here, and on the other hand the three last papers are not based on talks given at the seminar. The papers in this volume are arranged according to the order of their presentation at the seminar. The first paper of Bourgain and the note of Gromov are based on talks presented in previous years.

Joram Lindenstrauss, Vitali Milman

GFA

1985–1986

List of GAFA-Seminar Talks 1985-1986

25 October 85, Milman V.D., Diameter of a minimal invariant subset of equivariant Lipschitz actions on compact subsets of $\mathbb{R}^k$.

8 November 85, Lindenstrauss J., The weak distance between Banach spaces with a symmetric basis (joint work with A. Szankowski).

22 November 85, 1. Tzafriri L. Decomposition of unconditional bases in direct sums (joint work with P. Casazza and N. Kalton)

2. Tzafriri L., Cofactors of Euclidean subspaces of ℓ_p^n (joint work with J. Bourgain).

29 November 85, 1. Barany I., It is difficult to compute the volume of a convex body (joint work with Z. Füredi).

2. Barany I., Applications of Borsuk Theorem.

6 December 85, 1. Kalton N., Some remarks on r.i. Banach Lattices.

2. Kalton N., Topologies on Banach spaces generated by balls (joint work with G. Godefroy).

20 December 85, 1. Schechtman G., Packing k-dimensional subspaces of L_1 in ℓ_1^n.

2. Milman V.D., Some remarks related to Urysohn's Inequality.

27 December 85, Milman V.D., Inverse Brunn-Minkowsky Inequality and applications to Linear Theory of Normed Spaces.

10 January 86, 1. Schechtman G., On normed subspaces of L_0 (after N. Kalton).

2. Reisner S., A different proof of the volume characterization of Hansen Lima spaces by M. Meyer.

9 March 86, 1. Sternfeld Y., Uniform separation of points and measures.

2. Gordon Y., A recent remark of J.P. Kahane.

21 March 86, Schachermayer W., RNT and KMT are equivalent for strongly regular spaces.

4 April 86, 1. Milman V.D., Some new results of Pisier on projections and some new results on Entropy.

2. Arazy J., Banach spaces with biholomorphically equivalent unit balls are isometric (after W. Kaup and H. Upmeir).

20 April 86, Gromov M., Lipshitz geometry of expanding spaces.

9 May 86, 1. Schechtman G., A new approach to the problem of embedding into ℓ_p^n.

2. Bourgain J., Problems in harmonic analysis related to Sidon sets.

19 May 86, 1. Bourgain J., Remarks on the extension of Lipschitz maps defined on discrete sets.

2. Gordon Y., Some variants of Slepian's lemma.

3. Milman V.D., Generalization of the spherical isperimetric inequality to the uniformly convex Banach spaces (joint work with M. Gromov).

5 June 86, 1. Lewis D., The projection constant of an n-dimensional space is strictly smaller than $\sqrt{n}$ (joint work with H. König).

2. J. Bourgain, Maximal inequalities for convex bodies in $\mathbb{R}^n$.

Table of Contents

MONOTONICITY OF THE VOLUME OF INTERSECTION OF BALLS

M. Gromov

I.H.E.S.
France

1. Consider points $x_i \in \mathbb{R}^n$, for $i = 1, \dots, k$ and let $B(x_i, r_i) \subset \mathbb{R}^n$ be the balls around these points of given radii $r_i \geq 0$. Let

$$V(x_i, r_i) = \text{Vol} \bigcap_i B(x_i, r_i) \quad .$$

1.A THEOREM. *if $k \leq n+1$ then the function V is monotone decreasing in $d_{ij} = \|x_i - x_j\|$, that is if k-tuples x_i and x'_i have $d_{ij} \geq d'_{ij}$ then*

$$V(x_i, r_i) \leq V(x'_i, r_i) \quad .$$

1.B Take the sphere $S_r^{n-1} \subset \mathbb{R}_n$ of a fixed radius R around the origin and consider the spherical volume of the intersections of the balls B_i with S_r^{n-1},

$$V_{n-1}(x_i, r_i, r) = \text{Vol } S_r^{n-1} \bigcap_i B_i(x_i, r_i)$$

1.B′ LEMMA. *If $k \leq n$ and some k-tuples of points x_i and x'_i in $\mathbb{R}^n$ have $\|x_i\| = \|x'_i\|$ for $i = 1, \dots, k$ and $d_{ij} \geq d'_{ij}$, then*

$$V_{n-1}(x_i, r_i, r) \leq V(x'_i, r_i, r)$$

for all $r \geq 0$ and $r_i \geq 0$.

1.B″ REMARK: The most important case of 1.B′ is that where $\|x_i\| = \|x'_i\| = r$ which gives a version of 1.A for the sphere S_r^{n-1}. Notice that 1.B′ follows from this special case applied to the radial projection $\bar{x}_i$ of x_i to S_r^n and to $\bar{r}_i$, such that

$$B(\bar{x}_i, \bar{r}_i) \cap S_i^{n-1} = B(x_i, r_i) \cap S_r^{n-1} \quad .$$

Furthermore, since the geometry of S_r^{n-1} converges to that of $\mathbb{R}^{n-1}$ the spherical version of 1.A implies the Euclidean version.

1.C THE PROOF OF 1.B′: Assume, that 1.B′ by induction holds true for given k and $n-1$ and let us prove it for $(k+1)$-tuples in $S^n_r \subset \mathbb{R}^{n-1}$. By the above remark we may assume that the $(k+1)$-tuples in question, say $(x_0, \ldots, x_k)$ have $\|x_i\| = r$ for $i = 0, \ldots, k$. We also may assume for two $(k+1)$-tuples (x_i) and (x'_i) under comparison that $x_0 = x'_0$. Now we prove 1.B′ for S^n_r under an additional

(a) *Technical Assumption*

$$\|x_i - x_0\| = \|x'_i - x_0\| = r'_i \quad \text{for } i = 1, \ldots, k \quad .$$

The intersection in question is

$$S^n_r \cap B(x_0, r_0) \bigcap_{i=1}^{k} B(x_i, r_i) = \bigcup_{t=0}^{r} S^{n-1}(r_t) \bigcap_{i=1}^{k} B(x_i, r_i) \quad ,$$

where $S^{n-1}(r_t)$ is the sphere

$$S^{n-1}(r_t) = S^n_r \bigcap B(x_0, t) \quad .$$

By the induction assumption

$$\mathrm{Vol}_{n-1} S^{n-1}(r_t) \bigcap_{i=1}^{k} B(x_i, r_i) \leq \mathrm{Vol}_{n-1} S^{n-1}(r_t) \bigcap_{i=1}^{k} B(x'_i, r_i)$$

and the proof under assumption (a) follows by integration in t.

2. REDUCING THE GENERAL CASE TO THAT SATISFYING (A):

2.A. LEMMA. *Let x_i and x'_i in $\mathbb{R}^n$ for $i = 1, \ldots, k \leq n$ have $d_{ij} \geq d'_{ij}$ then there exists a continuous family of k-tuples x^t_i, such that $x^0_i = x_i$ and $x^1_i = x'_i$ and d^t_{ij} is decreasing in t.*

PROOF: The mutual distances d_{ij} can be replaced by scalar products $\langle x_i, x_j\rangle$ and the pertinent deformation is the linear homotopy between the matrices $\langle x_i, x_j\rangle$ and $\langle x'_i, x'_j\rangle$. Since the matrix $(1-t)\langle x_i, x_j\rangle + t\langle x'_i, x'_j\rangle$ is positive semidefinite for all $t \in [0,1]$ (as well as $\langle x_i, x_j\rangle$ and $\langle x'_i, x'_j\rangle$) it represents some points x^t_i with the mutual distances d^t_{ij}. Q.E.D.

CONCLUSION OF THE PROOF:

Any monotone (for d^t_{ij}) homotopy can be approximated by another homotopy which also is (nonstrictly) monotone and which, on every subsegment in $[0,1]$ of the form $[\frac{\nu}{N}, \frac{\nu+1}{N}]$ (for N depending on desired the precision of approximation) has all but one among d^t_{ij} constant in t.

Now the special case applies to the tuples x_i^t and $x_i^{t'}$ for $t = \frac{\nu}{N}$ and $t' = \frac{\nu+1}{N}$ for $\nu = 1, \ldots, N$ and the proof in the general case is concluded.

REMARK: It is likely that 1.A extends to the hyperbolic space with curvature $\equiv -$const. (as well as to the spheres S_r^n with curvature $\equiv +\text{const} = r^{-2}$). The above proof breaks down at the only 'non-trivial'point that is Lemma 2.A.

3. COROLLARIES.

3.A (Kirszbraun Intersection Property; see [W.W]) if x_i and x_i' for $i = 1, \ldots, k$, in $\mathbb{R}^n$, now for any k, have $d_{ij} \geq d'_{ij}$, then

$$\bigcap_i B(x_i, r_i) \neq \emptyset \Rightarrow \bigcap_i B(x_i', r_i) \neq \emptyset \quad .$$

PROOF: Apply 1.A. to the balls around x_i and x_i' in the ambient space $\mathbb{R}^m \supset \mathbb{R}^n$ for $m = \max(n, k-1)$.

3.B COROLLARY. *(Kirszbraun; see [W.W]) Let X and Y be (finite or infinite dimensional) Hilbert spaces and set $f : X_0 \to Y$ be a distance decreasing map of a subset $X_0 \subset X$. Then f extends to a distance decreasing map $X \to Y$. Furthermore, a similar result holds for maps of subsets in the hemisphere of the Hilbert sphere.*

3.C. *Let $e(r)$ be a monotone decreasing function in $r \in [0, \infty)$ such that*

$$\int_0^\infty |e(r)| < \infty \quad .$$

Assign to each $x \in \mathbb{R}^n$ the function $e_x = e_x(y) = e(|x - y|)$.

3.D SLEPIAN INEQUALITY. *[S] For the above x_i and x_i'*

$$\int_{\mathbb{R}^n} \prod_i e_{x_i} \leq \int_{\mathbb{R}^n} \prod_i e_{x_i'} \quad .$$

The proof obviously follows from 1.A.

QUESTION: Who is the author of 1.A.? My guess is this was known to Archimedes. Undoubtedly the theorem can be located (in the form 3.D as well as 1.A) somewhere in the 17th century.

REFERENCES

[S] Slepian D. The one-sided barrier problem for Gaussian noise, Bell System Tech. J. 41 (1962), 463-501.

[W.W] Wells J.H. and Williams L.R. Embeddings and Extensions in Analysis, Ergebuisse n.84, Springer Verlag 1975.

ON LATTICE PACKING OF CONVEX SYMMETRIC SETS IN $\Re^n$

J. BOURGAIN
IHES
University of Illinois

1. Introduction

Let $n \geq 2$ be a positive integer. A lattice L on $\Re^n$ is determined by n linearly independent vectors $e_1, \ldots, e_n$ in $\Re^n$;

$$L = \{u_1 e_1 + \cdots + u_n e_n \mid u_k \in Z \quad (1 \leq k \leq n)\}.$$

Let $\| \ \|$ be a norm on $\Re^n$ and $C = \{x \in \Re^n \mid \|x\| \leq 1\}$ the corresponding unit ball. Define following numbers

$$d_1(C,L) = \frac{1}{2} \inf_{x,y \in L, x \neq y} \|x-y\| = \frac{1}{2} \inf_{x \in L, x \neq 0} \|x\|$$

$$d_2(C,L) = \sup_{x \in \Re^n} \inf_{y \in L} \|x-y\|.$$

Suppose $x \in L, x \neq 0$ satisfies $d_1(C,L) = \frac{1}{2}\|x\|$. Then if $y \in L$

$$\left\|\frac{x}{2} - y\right\| \geq \|x-y\| - \frac{1}{2}\|x\| \geq d_1(C,L).$$

Hence $d_1(C,L) \leq d_2(C,L)$.
Define

$$\rho(C) = \inf \frac{d_2(C,L)}{d_1(C,L)} \geq 1$$

where the infimum is taken over all lattices L on $\Re^n$.
If we let $L = Z^n \subset \Re^n$, then $d_1(C,L) = d_2(C,L) = \frac{1}{2}$ if $\|x\| = \max |x_i|$ while $d_2(C,L) \sim \sqrt{n}\, d_1(C,L)$ for $\|x\| = (\Sigma x_i^2)^{1/2}$ the euclidean norm. The main result of this paper is the following fact

Proposition 1: $\rho(C)$ is bounded by an absolute constant, independent of the dimension and the norm.
Equivalently

Proposition 1': There is an absolute constant γ such that given a convex symmetric body C in $\Re^n$, there exists a lattice L satisfying the conditions

$$(u+\gamma C) \cap (u'+\gamma C) = \phi \quad \text{if } u \neq u' \text{ in } L \tag{1}$$

$$\Re^n = \bigcup_{u \in L} (u+C). \tag{2}$$

The statement in the particular case of the euclidean ball answers a question raised by M. Gromov related to the construction of almost flat manifolds [2].

The norm $\| \ \|$ is called uniformly convex provided there is a function $\delta :]0,1] \rightarrow]0,1]$

satisfying

$$\|x\| = 1 = \|y\| \quad \text{and} \quad \|x-y\| > \varepsilon \Rightarrow \left\|\frac{x+y}{2}\right\| < 1 - \delta(\varepsilon).$$

Proposition 2: If C is uniformly convex, then $\rho(C) > 1 + \tau$ where $\tau > 0$ depends only on the convexity modulus δ.
Actually, Proposition 2 follows immediately from an improved version of Minkowski's first theorem for uniformly convex bodies.

Proposition 3: Let λ_1 be the first minimum of C, i.e.

$$\lambda_1 = \inf\{t > 0 \mid (tC) \cap Z^n \neq \{0\}\}.$$

Then

$$\lambda_1^n \cdot Vol\ C \leq (2-\tau)^n$$

where again τ depends on the convexity modulus.
The argument presented here gives $\rho(C) \leq 16^{1/n} 4$ in the general case. As will be clear, this bound is certainly not best possible.

Remark: Recently, the paper [B] was brought to my attention. The result of [B] encludes Propositions 1 and 1' and also Remark 3 at the end. The present exposition is simpler however, both technically and conceptually. Our method also enables an extension to finite families since it is purely random (see Remark 4).

2. Proof of Proposition 1

The proof is based on an averaging argument, involving the "random" lattice $\Lambda_{z,\alpha}$ determined by the columns of the matrix

$$\begin{bmatrix} \alpha^{-\frac{1}{n-1}} & 0 & \cdots & 0 & z_1\alpha^{-\frac{1}{n-1}} \\ 0 & \alpha^{-\frac{1}{n-1}} & \cdots & 0 & z_2\alpha^{-\frac{1}{n-1}} \\ \vdots & \vdots & & & \vdots \\ 0 & 0 & \cdots & \alpha^{-\frac{1}{n-1}} & z_{n-1}\alpha^{-\frac{1}{n-1}} \\ 0 & 0 & \cdots & 0 & \alpha \end{bmatrix}$$

where $z = (z_1, \ldots, z_{n-1}) \in [0,1]^{n-1}$ and $\alpha > 0$. Note that $\det \Lambda_{z,\alpha} = 1$. These lattices appear in the proof of the Davenport-Rogers lemma (see [3], p. 176) which may be stated as follows.

Lemma 1: If χ is a bounded Riemann integrable function on $\Re^n$ with compact support, then

$$\lim_{\alpha\to 0} \int_{[0,1]^{n-1}} \{ \sum_{\substack{x\neq 0 \\ x\in\Lambda_{z,\alpha}}} \chi(x)\} dz = \int_{\Re^n} \chi(x)dx. \tag{3}$$

For completeness' sake, the argument will be included in the next section. This lemma will permit us to obtain (1) in Proposition 1'. Condition (2) will be derived from the following fact:

Lemma 2: Let f be a compactly supported function on $\Re^n$ with Fourier transform $\hat{f}$ in $L^1(\Re^n)$. Then

$$\overline{\lim_{\alpha\to 0}} \int_{[0,1]^{n-1}} \sup_{y\in\Re^n} |\int f - \sum_{x\in\Lambda_{z,\alpha}} f(y-x)| dz \le \|\hat{f}\|_1. \tag{4}$$

We postpone the proof of (3), (4) to the next section.

Proof of Proposition 1'

Without restriction, we may assume $Vol\ C = 2^{n+1}$. Define

$$t = (8\ Vol C)^{-1/n}$$

$$\chi = \chi_{tC} \quad \text{(indicator function)}$$

$$f = 4^n(\chi_{\frac{1}{2}C} * \chi_{\frac{1}{2}C}).$$

Hence

$$\int\chi = t^n\,Vol\ C$$

$$\int f = (Vol\ C)^2$$

$$\|\hat{f}\|_1 = \int |\hat{\chi}_C(\tfrac{1}{2}\xi)|^2 d\xi = 2^n\,Vol\ C \quad \text{by Parseval's identity.}$$

Let $\rho = 2(Vol\ C)^2$. Apply (3) and (4) with α small enough and multiply both members of (3) by the factor ρ. There follows the existence of a point $z \in [0,1]^{n-1}$ and hence a lattice $L = \Lambda_{z,\alpha}$ satisfying

$$\rho \sum_{x\in L-0} \chi_{tC}(x) + \sup_{y\in\Re^n} |(Vol\ C)^2 - \sum_{x\in\Lambda_{z,\alpha}} f(y-x)| \le (2^n+\rho t^n)\,Vol\ C + 1. \tag{5}$$

Hence

$$\sum_{x\in L-0} \chi_{tC}(x) \le (2^n\rho^{-1}+t^n)\,Vol\ C + \frac{1}{5} < 1 \Rightarrow \sum_{x\in L-0} \chi_{tC}(x) = 0.$$

Thus (1) of Proposition 1' holds with

$$\gamma = \frac{t}{2} = \frac{1}{4 2^{4/n}}.$$

(5) also implies that whenever $y \in \Re^n$

$$\sum_{x\in\Lambda_{z,\alpha}} f(y-x) \ge (Vol\ C)^2 - (2^n+\rho t^n)\,Vol\ C - 1 = \frac{1}{4}(Vol\ C)^2 - 1 > 0$$

and consequently, for some $x \in L$,

$$y \in x + supp\ f \subset x + C$$

by definition of F and convexity of C. This yields (2). Note that $\rho(C) \leq \frac{1}{\gamma}$ and hence $\rho(C) \leq 4.16^{1/n}$.

3. Proof of the Lemmas

Proof of Lemma 1: If $x = (x_1, \ldots, x_{n-1}, 0) \in A_{z,\alpha}$, $x \neq 0$, it follows that $|x| \geq \alpha^{-1/n-1}$ and thus $\chi(x) = 0$ if α small enough. Hence, for sufficiently small α, we may write

$$\sum_{\substack{x \neq 0 \\ x \in \Lambda_{z,\alpha}}} \chi(x) =$$

$$= \sum_{\substack{r \neq 0 \\ r \in Z}} \sum_{u_1, \ldots, u_{n-1} \in Z} \chi(\alpha^{-1/n-1}(u_1 + rz_1), \ldots, \alpha^{-1/n-1}(u_{n-1} + rz_{n-1}), \alpha r).$$

Exploiting $\sum_{u \in Z^{n-1}}$, it follows

$$\int_{[0,1]^{n-1}} \{ \sum_{\substack{x \neq 0 \\ x \in \Lambda_{z,\alpha}}} \chi(x)\} dz_1 \cdots dz_{n-1} =$$

$$\sum_{\substack{r \neq 0 \\ r \in Z}} \sum_{u \in Z^{n-1}} r^{-n+1} \int_{[0,r]^{n-1}} \chi(\alpha^{-1/n-1}(u_1 + z_1), \ldots, \alpha^{-1/n-1}(u_{n-1} + z_{n-1}), \alpha r) dz =$$

$$\sum_{\substack{r \neq 0 \\ r \in Z}} \int_{-\infty}^{\infty} \cdots \int_{-\infty}^{\infty} \chi(\alpha^{-1/n-1} y_1, \ldots, \alpha^{-1/n-1} y_{n-1}, \alpha r) dy_1 \cdots dy_{n-1} =$$

$$\alpha \sum_{\substack{r \neq 0 \\ r \in Z}} \int_{-\infty}^{\infty} \cdots \int_{-\infty}^{\infty} \chi(y_1, \ldots, y_{n-1}, \alpha r) dy_1, \ldots, dy_{n-1}.$$

Passing to the limit with $\alpha \to 0$, the latter sum converges indeed to $\int \chi$.

Proof of Lemma 2: Clearly

$$\sum_{x \in \Lambda_{z,\alpha}} f(y - x) = = \lim_{R \to \infty} \sum_{|r| \leq R} \frac{R - |r|}{R} \sum_{u \in Z^{n-1}}$$

$$f(y_1 - \alpha^{-1/n-1}(u_1 + rz_1), \ldots, y_{n-1} - \alpha^{-1/n-1}(u_{n-1} + rz_{n-1}), y_n - \alpha r).$$

Define for $x' = (x_1, \ldots, x_{n-1})$ and fixed α

$$g_\alpha(x', x_r) = f(\alpha^{-1/n-1} x', \alpha x_n)$$

$$G_\alpha(x', x_n) = \sum_{u \in Z^{n-1}} g_\alpha(x' + u, x_n).$$

Taking the supremum over $y \in \Re^n$, we may replace y by $(\alpha^{-1/n-1} y_1, \ldots, \alpha^{-1/n-1} y_{n-1}, \alpha y_n)$. Hence, (4) will result from

$$\overline{\lim_{\alpha\to 0}} \int_{[0,1]^{n-1}} \sup_{y\in\Re^n} \overline{\lim_{R\to\infty}}$$

$$|\int f - \sum_{|r|\le R} \frac{R-|r|}{R} G_\alpha(y_1-rz_1,\ldots,y_{n-1}-rz_{n-1},y_n-r)|\,dz \le \|\hat f\|_1. \tag{6}$$

The function G_α is periodic in the first $(n-1)$-variables and will be considered as a function on $(\Re^{n-1}/Z_{n-1})\times\Re$. Fourier expansion yields

$$G_\alpha(x',x_n) = \int \sum_{J\in Z^{n-1}} \hat g_\alpha(J,\lambda)e^{2\pi i(<J,x'>+\lambda x_n)}d\lambda.$$

Hence

$$\sum_{|r|\le R} \frac{R-|r|}{R} G_\alpha(y_1-rz_1,\ldots,y_{n-1}-rz_{n-1},y_n-r) =$$

$$\int\{\sum_{J\in Z^{n-1}} \hat g_\alpha(J,\lambda)e^{2\pi i(<J,y>+\lambda y_n)}F_R(<J,z>+\lambda)\}d\lambda$$

where $y=(y_1,\ldots,y_{n-1})$ and $F_R(t) = \sum_{|r|\le R}\frac{R-|r|}{R}e^{2\pi irt}$ stands for the Fejer kernel. Estimate the left member of (6) as

$$\overline{\lim_{\alpha>0}} \sup_{y\in\Re^n} \overline{\lim_{R\to\infty}} |\int f - \sum_{|r|\le R}\frac{R-|r|}{R}\int \hat g_\alpha(0,\lambda)e^{2\pi i\lambda(y_n-r)}d\lambda| + \tag{7}$$

$$\overline{\lim_{\alpha>0}} \sup_R \int_{[0,1]^{n-1}} \int \sum_{\substack{J\in Z^{n-1}\\ J\ne 0}} |\hat g_\alpha(J,\lambda)| F_R(<J,z>+\lambda)\,d\lambda\, dz. \tag{8}$$

Estimation of (7): Clearly

$$\int \hat g_\alpha(0,\lambda)e^{2\pi i\lambda(y_n-r)}d\lambda = \int g_\alpha(x',y_n-r)dx' = \alpha\int f(x',\alpha(y_n-r))dx'.$$

Thus

$$\lim_{R\to\infty} |\int f - \sum_{|r|\le R}\frac{R-|r|}{R}\int \hat g_\alpha(0,\lambda)e^{2\pi i\lambda(y_n-r)}d\lambda| = |\int f - \alpha\sum_{r\in Z}\int f(x',\alpha y_n+\alpha r)dx'| \le$$

$$\sup_{|s|<\alpha,\, t\in\Re} \int |f(x',t+s)-f(x',t)|\,dx' \xrightarrow{\alpha\to} 0.$$

Estimation of (8): Since $\int F_r(t)dt = 1$, we get

$$\overline{\lim_{\alpha\to 0}} \int \sum_J |\hat g_\alpha(J,\lambda)|\,d\lambda =$$

$$\overline{\lim_{\alpha\to 0}} \int \sum_J |\hat f(\alpha^{1/n-1}J,\alpha^{-1}\lambda)|\,d\lambda =$$

$$\overline{\lim_{\alpha\to 0}} \sum_J \alpha \int |\hat f(\alpha^{1/n-1}J,\lambda)|\,d\lambda = \|\hat f\|_1.$$

4. Proof of Proposition 3

Let C be the unit ball of a uniformly convex norm $\| \ \|$, $C = \{x \in \Re^n ; \|x\| \le 1\}$ and let δ be the convexity modulus of $\| \ \|$. The proof will follow from the simple observation that one may replace C by a convex symmetric $\bar{C}$ such that $\lambda_1(\bar{C}) \ge \lambda_1(C)$ and $(Vol\ \bar{C})^{1/n} \ge (1+\gamma)(Vol\ C)^{1/n}$.

Lemma 3: Let $0 \le \eta < 1$ and $x \in [(1+\eta)C] \backslash C$. Then there is a convex symmetric D contained in $(1+\eta)C$ such that $x \notin D$ and

$$Vol[(1+\eta)C] - Vol\ D < [2\delta^{-1}(\eta)]^n\ Vol\ C$$

where δ^{-1} stands for the inverse function of δ.

Proof: A separation argument yields a functional f such that

$$f(x) > \|f\| = \sup_{y \in C} |f(y)| = 1.$$

Define $D = [(1+\eta)C] \cap [|f| \le 1]$. If $\|y\|, \|z\| \le 1+\eta$ and $f(y), f(z) > 1$, it follows that

$$1 < f(\frac{y+z}{2}) \le \|\frac{y+z}{2}\| < (1+\eta)[1-\delta(\frac{\|y-z\|}{1+\eta})]$$

$$\delta(\frac{\|y-z\|}{1+\eta} < \frac{\eta}{1+\eta} \Rightarrow \|y-z\| < 2\delta^{-1}(\eta).$$

Consequently

$$Vol\ [(1+\eta)C] - Vol\ D < [2\delta^{-1}(\eta)]^n\ .\ Vol\ C.$$

Proof of Proposition 3: Assume $\lambda_1(C) = 1$. Leave η to be specified later. If $x \in Z^n \cap (1+\eta)C, x \ne 0$, then we may apply the previous lemma. Also since $\|x-y\| \ge 1$ for $x \ne y$ in Z^n, a simple volume consideration shows that

$$\#\ [Z^n \cap (1+\eta)C] \le (3+2\eta)^n < 5^n.$$

Intersecting the bodies D given by Lemma 3 yields $\bar{C} \subset (1+\eta)C$ and satisfying the conditions

$$\lambda_1(\bar{C}) \ge 1$$

$$Vol\ \bar{C} \ge \{(1+\eta)^n - [10\delta^{-1}(\eta)]^n\}\ Vol\ C.$$

If we let $\eta = \delta(\frac{1}{10})$, application of Minkowski's first theorem yields

$$2^n \ge (Vol\ \bar{C})\lambda_1(\bar{C})^n \ge \delta(\frac{1}{10})[1+\delta(\frac{1}{10})]^{n-1}\lambda_1(C)^n\ Vol\ C$$

and the proposition follows.

In the case of ℓ^p-norm on $\Re^n$, thus $\|x\| = (\sum_{i=1}^{n} |x_i|^p)^{1/p}$, we have $\delta(\frac{1}{10}) \sim \frac{1}{p+p'}$, $p' = \frac{p}{p-1}$. Hence, in view of the volume expression for the ℓ_n^p-unit ball (see [6], exp. CIII).

Corollary 4: Let $A = (a_{ij})$ be an $(n \times n)$-matrix. If

$$(det\ A)^{1/n} < (1+\frac{c}{p+p'})(pe)^{1/p}\,\Gamma(1+\frac{1}{p})n^{-1/\max(2,p)} \qquad (*)$$

there is a point $x \in Z^n$, $x \neq 0$ satisfying

$$\Sigma_i \mid \Sigma_j a_{ij} x_j \mid^p \leq 1.$$

Here c is a constant and n assumed sufficiently large.

The improvement of this estimate with respect to Minkowski's theorem is the presence of the factor $1 + \frac{c}{p+p'}$, in the right member of (*). Already for the euclidean norm, the sharp result does not seem to be known. Applying the result of Blichtfeldt (see [3], p. 295) for ellipsoids

$$\lambda_1^n\ Vol\ \mathcal{E} \leq \frac{n+2}{2} 2^{n/2}$$

it follows that for $p = 2$, the factor $1 + \frac{c}{p+p'}$ in (*) may be replaced by any number less than $\sqrt{2}$. This may be slightly improved, due to subsequent work (up to $2^{0,599}$) (see [7], p. 321 for a survey).

5. Remark and Problems

(1) The consecutive minima of the convex symmetric body C are defined by

$$\lambda_i = \inf\{\lambda > 0 \mid \dim\ (\lambda C \cap Z^n) \geq i\} \quad (1 \leq i \leq n).$$

Minkowski's second theorem asserts that

$$\lambda_1 \cdots \lambda_n\ Vol\ C \leq 2^n.$$

For uniformly convex C, there is again an improvement

$$\lambda_1 \cdots \lambda_n\ Vol\ C \leq (2-\tau)^n$$

with $\tau > 0$ dependent on the convexity modulus of C only. The proof is in the spirit of Proposition 3. Using a change of coordinates by an element of the group $\pm SL(Z^n)$, we may assume

$$x \in \lambda C \cap Z^n, \lambda < \lambda_i \Rightarrow x_i = \cdots = x_n = 0 \qquad (**)$$

(see [1], Appendix B). Next, we replace C by a convex symmetric $D \subset (1+\eta)C$

$$Vol\ D \cong Vol\ [(1+\eta)C] \qquad (***)$$

and such that D still satisfies (**) for $i \leq \alpha\eta$, where $\alpha > 0$ depends on the convexity modulus. This is achieved by slicing away a small part of $(1+\eta)C$ just as in proving Proposition 3. In this case, some additional entropy considerations in k-dimensional spaces ($k \leq \alpha n$) are involved. We let the reader work out the details. Next, apply Minkowski's second theorem to D

$$\lambda_1(D) \cdots \lambda_n(D)\ Vol\ D \leq 2^n.$$

But, by construction and (**)

$$\lambda_i(D) \geq \lambda_i \quad \text{for } i \leq \alpha n,\ \lambda_i(D) \geq (1+\eta)^{-1}\lambda_i \quad \text{in general.}$$

Thus, by (***)

$$\lambda_1 \cdots \lambda_n (1+\eta)^{\alpha n}\ Vol\ C \leq \frac{1}{2} 2^n$$

whence the conclusion.

(2) The previous argument makes strong use of uniform convexity. It may be of interest to understand the effect of parameters such as type, cotype (see [4]) with respect to Minkowski's inequality. Since equality appears only for affine images of the cube, it may be expected that a cotype hypothesis leads to a substantial improvement of the general result.

(3) Proposition 1' remains valid if C is only assumed convex, replacing γ by $\frac{\gamma}{2}$. We use a result due to C. Rogers and G. Shephard [5]

$$Vol\ (C-C) \leq \binom{2n}{n}\ Vol\ C < 4^n\ Vol\ C$$

where

$$C - C = \{x-y \mid x,y \in C\}.$$

In the proof of Proposition 1', replace t by $\frac{1}{4}(8\ Vol\ C)^{-1/n}$ and χ by $\chi_{t(c-c)}$. The property

$$\sum_{x \in L-0} \chi_{t(c,c)}(x) = 0$$

then implies (2).

(4) From the argument, it is clear that the lattice L in Proposition 1' may be constructed for a finite family of convex bodies C (modifying the value of γ).

REFERENCES

[1] J.W.S. Cassels: An introduction to diophantine approximation, Cambridge UP, 1965.

[2] M. Gromov, Private communication.

[3] C. G. Lekkerkerker, Geometry of numbers, North-Holland, Amsterdam, 1969.

[4] J. Lindenstrauss, L. Tzafriri: Classical Banach spaces II, Springer 17, 1979.

[5] C. Rogers. G. Shephard, The difference body of a convex body, Arch Math. 8 1957, 220-223.

[6] M. Rogalski, Sur le quotient volumique d'un espace de dimension finie, Publication Math. de l'Universite Pierre et Marie Curie, N. 46, 1980/81.

[7] G. Fejes Toth, New results in the theory of packing and covering, in "convexity and its applications", Birkhauser Verlag 1983.

[8] G. J. Butler, Simultaneous packing and covering in euclidean space, Proc. London. Math. Soc., 25, 1972, 721-735.

DIAMETER OF A MINIMAL INVARIANT SUBSET OF EQUIVARIANT LIPSCHITZ ACTIONS ON COMPACT SUBSETS OF $\mathbb{R}^k$

V.D. Milman*
I.H.E.S and
School of Mathematical Sciences
Raymond and Beverly Sackler
Faculty of Exact Sciences
Tel Aviv University
Tel Aviv, Israel

1. Introduction.

The main idea of this note originates in [GM] (a part of that paper which deals with fixed point theorems). In the first part (sections 2-4) of our note we essentially repeat some of the results from [GM] with a different interpretation of results and proofs. Also new examples will be considered. In the second part (section 5) we present a finite dimensional version of the results from section 4. In this case we don't have any more fixed point theorems; however, we may estimate from above a diameter of a minimal invariant subset and this is the main purpose of the note.

Also note that all results are heavily based on a concentration of measure property of various classes of groups and homogeneous spaces - the so called Levy families which we define in section 3 (see [GM], [AM] or [MSch]).

A finite dimensional interpretation of some results from [GM] in a different direction may be found in [G], §9.

2. G-space; concentration property.

Let M be a metric space (with the metric ρ) and let G be a family of uniformly equicontinuous maps from M to M. This means that there exists a continuous function $\omega(\varepsilon)$ (defined for $\varepsilon \geq 0$), $\omega(0) = 0$, such that $\rho(gx, gy) \leq \omega\big(\rho(x,y)\big)$ for any $x, y \in M$ and any $g \in G$. Note that we don't assume that G is a group (or even a semigroup) although it will often be the case.

Definition 2.1. We say that a subset $A \subset M$ is essential (with respect to the action of G) iff for every $\varepsilon > 0$ and every finite subset $\{g_1, \ldots, g_n\} \subset G$

$$\bigcap_{i=1}^{n} g_i A_\varepsilon \neq \emptyset$$

* Supported in part by the US-Israel Binational Science Foundation

where $A_\varepsilon = \{x \in M : \rho(x, A) \leq \varepsilon\}$.

Definition 2.2. The pair (M, G) has the *property of concentration* iff for every finite covering $M = \bigcup_1^N M_i$, $M_i \subset M$, there exists M_{i_0} which is essential (for the action of G).

2.3 Examples. A. Let H be an infinite dimensional Hilbert space and $\{e_i\}_{i=1}^\infty$ an orthonormal basis of H. Let the orthogonal group $SO(n)$ be realized as unitary operators on H which are the identity on $\text{span}\{e_i\}_{i>n}$. Then $SO(n) \subset SO(n+1)$, $n \in \mathbb{N}$. We consider $M = G = SO(\infty) \stackrel{\text{def}}{=} \bigcup_{n\geq 1} SO(n)$ with the Hilbert-Schmidt operator metrip ρ on M. Then:

(2.3.A) $\qquad (M, G)$ *has the concentration property .*

(We will prove this in the next section.)

B. Let $S^\infty = \{x \in H; |x| = 1\}$ be the unit sphere of the Hilbert space H with the standard euclidean distance. Then

(2.3.B) $\qquad (S^\infty, G = SO(\infty))$ *has the concentration property .*

C. Let u be any unitary operator on H and $G = \{u^n\}_{n=-\infty}^\infty$. Then

(2.3.C) $\qquad (S^\infty, G)$ *has the concentration property*

(see proof in the next section).

D. **Generalization of C.** Let $\mathcal{M}$ be a family of pairwise commuting unitary operators in H. Then

(2.3.D) $\qquad (S^\infty, \mathcal{M})$ *had the concentration property*

(proof is the same as for (2.3.C)).

It will be clear from the proofs below that every Levy family of spaces (see definition in section 3) corresponds to some space with the concentration property in the above sense. Many examples of Levy families may be found in [GM], [AM] or [MSch]. The following problem originated from [GM] (however it was not stated there).

2.4. Problem. Let U be the group of *all* unitary operators $H \to H$, $\dim H = \infty$. Does (S^∞, U) have the concentration property?

3. Levy Families.

Let (X_n, ρ_n, μ_n), $n = 1, 2, \ldots$, be a family of metric spaces (X_n, ρ_n) which are also probability spaces with probability measures μ_n defined on the Borel subsets of X_n. For $A \subset X_n$ and $\varepsilon > 0$ we define $A_\varepsilon = \{x \in X_n; \rho_n(x, A) \leq \varepsilon\}$. Let diam $X_n = d_n$ and $d_n \geq 1$. We define the *concentration function* of a space (M, ρ, μ)

$$\alpha(M, \varepsilon) = 1 - \inf\{\mu(A_\varepsilon), \text{ for any Borel set } A \subset X_n \text{ such that } \mu(A) \geq 1/2\} .$$

Definition 3.1. The family (X_n, ρ_n, μ_n), $n \in I\!N$, is said to be a *Levy family* if for every $\varepsilon > 0$

$$\alpha(X_n; \varepsilon d_n) \to 0 \qquad (n \to \infty) ,$$

and it is said to be a *normal Levy family* with constants C_1 and C_2 if

$$\alpha(X_n; \varepsilon d_n) \leq C_1 \exp(-C_2 \varepsilon^2 n) .$$

In this note we use the following examples of Levy families.

i) The euclidean spheres (S^n, ρ, μ) equipped with the geodesic distance ρ (or, equivalently, with the norm distance $\rho(x, y) = |x - y|$) and the normalized (i.e., probabilistic) rotation invariant measure μ_n:

(3.1.i)
$$\alpha(S^{n+1}; \varepsilon) \leq \sqrt{\frac{\pi}{8}} \exp(-\varepsilon^2 n/2)$$

([L]; see also, e.g., [FLM]).

ii) The family $\{SO(n)\}_{n \in I\!N}$ equipped with the Hilbert-Schmidt operator metric ρ and the normalized Haar measure μ_n:

(3.1.ii)
$$\alpha(SO(n); \varepsilon) \leq \sqrt{\frac{\pi}{8}} \exp(-\varepsilon^2 n/8)$$

(see [GM]).

iii) If $E_2^n = \{-1, 1\}^n$ has the normalized Hamming metric

$$d(s, t) = \frac{1}{n} |\{i : s_i \neq t_i\}|$$

and the normalized counting measure, then

$$\alpha(E_2^n; \varepsilon) \leq \frac{1}{2} \exp(-2\varepsilon^2 n)$$

(see [AM]).

iv) The group Π_n of permutations of $\{1, \ldots, n\}$ with the normalized Hamming metric

$$d(\pi_1, \pi_2) = \frac{1}{n}|\{i : \pi_1(i) \neq \pi_2(i)\}|$$

and the normalized counting measure:

$$\alpha(\Pi_n; \varepsilon) \leq \exp(-\varepsilon^2 n/64) \ .$$

(B. Maurey [M]; see also [Sch] for a number of interesting generalizations). We also use the following easy consequence from the definition of $\alpha(M, \varepsilon)$.

Lemma 3.2. If $A \subset M$ *such that* $\mu(A) > \alpha(M, \varepsilon/2)$ *then* $\mu(A_\varepsilon) \geq 1 - \alpha(M, \varepsilon/2)$.

Proof. Consider the set $B = (A_\varepsilon)^c$ $(\equiv M \backslash A_\varepsilon)$. If $\mu(B_{\varepsilon/2}) \geq 1/2$ then $\mu(B_\varepsilon) \geq 1 - \alpha(M, \varepsilon/2)$ and $\mu(A) \leq \mu(B_\varepsilon^c) \leq \alpha(M, \varepsilon/2)$ which is impossible. Therefore $\mu(B_{\varepsilon/2}) < 1/2$ and $\mu[(B_{\varepsilon/2})^c_{\varepsilon/2}] \geq 1 - \alpha(M, \varepsilon/2)$. It remains to note that $(B_{\varepsilon/2})^c_{\varepsilon/2} \subset B^c = A_\varepsilon$.

3.3 Proof of (2.3.A). Let μ_n be the normalized Haar measure on $SO(n)$.

3.3.1. Let $A \subset SO(\infty)$ $(= \cup SO(n)$ as in 2.3) and for every $\varepsilon > 0$ $\overline{\lim_{n \to \infty}} \mu_n(A_\varepsilon \cap SO(n)) = 1$. Then A is an essential subset. (Proof is obvious.)

3.3.2. Let $SO(\infty) = \bigcup_{i=1}^{N} A_i$. Then there exists i_0 such that $\overline{\lim}\ \mu_n\left(\overline{A_{i_0} \cap SO(n)}\right) \geq 1/N$. By (3.1) and Lemma 3.2 it follows that for any $\varepsilon > 0$ the set $(A_{i_0})_\varepsilon$ satisfies 3.3.1. This end the proof of (2.3.A). □

Note that the proof of (2.3.B) is completely similar to the above one.

3.4. Proof of (2.3.C). By the spectral theory for unitary operators, we may find subspaces E_N, $\dim E_N = N \to \infty$, such that

$$\rho\big(u\ S(E_N), S(E_N)\big) < \delta_N \to 0 \quad \text{for} \quad N \to \infty$$

(here $S(E_N)$ is the unit sphere of the subspace E_N).

Let $S^\infty = \bigcup_{i=1}^{t} A_i$. Choose A_{i_0} such that for some infinite sequence $N_j (\to \infty)$

$$\mu_{S(E_N)}\left(\overline{A_{i_0} \cap S(E_N)}\right) \geq 1/t \quad \text{for} \quad N \in \{N_j\} \ .$$

We will prove that A_{i_0} is an essential subset of S^∞. Fix $\varepsilon > o$ and an integer T. We may choose N_0 such that for $N > N_0$, δ_N are so small that

$$\rho\big(u^k S(E_N), S(E_N)\big) < \varepsilon/3 \ , \quad k = 1, \ldots, T; N > N_0 \ .$$

Then for $N \in \{N_j\}$, $N > N_0$ and every $k = 0, 1, \ldots, T$

(3.2.) $$\mu_{S(E_N)}\big(u^k(A_{i_0})_{2\varepsilon/3} \cap S(E_N)\big) \geq 1/t\ .$$

Now we assume N_0 to be also so large that from (3.2) would follow by (3.1) and Lemma 3.2

$$\mu_{S(E_N)}\big(u^k(A_{i_0})_\varepsilon \cap S(E_N)\big) > 1 - 1/T$$

for every $k = 0, 1, , \ldots, T$ and $N \in \{N_j\}$, $N > N_0$. It is clear that $\bigcap_{k=0}^{T} u^k(A_{i_0})_\varepsilon \neq \emptyset$. □

4. Fixed point Theorems

In this section we *always* assume that M is a metric space having the concentration property (see definition 2.2) with respect to some action of G on M and K is a compact space.

Proposition 4.1. *Let $\varphi : M \to K$ be any map,* Dom $\varphi = M$*. Then there exists $x_0 \in K$ such that for any neighborhood $N(x_0)$ the set $\varphi^{-1}(N(x_0))$ is essential in M.*

(We call such point x_0 an *essential point of the map* φ).

Proof. If such x_0 does not exist then for any $x \in K$ we take an open neighborhood $0_x = N(x)$ of x such that $\varphi^{-1}(N(x))$ is not essential. So, we have the covering of K. Choose a finite subcovering $\{0_{x_i}\}_{i=1}^t$. Then $\cup\varphi^{-1}(0_{x_i}) = M$ and, by the concentration property, one of the sets $\varphi^{-1}(0_{x_i})$ has to be an essential set which contradicts the above construction. □

A map $\varphi : M \to K$ is called uniformly continuous if for any closed subset $A \subset K$ and any open neighborhood $0(A)$ there exists $\varepsilon > 0$ such that $\varphi([\varphi^{-1}(A)]_\varepsilon) \subset 0(A)$.

Theorem 4.2. *Let $\varphi : M \to K$ and $g : M \to M$, $g \in G$, be uniformly continuous maps and let a continuous map $A : K \to K$ be such that $\varphi g = A\varphi$. Then any essential point x_0 of φ is a fixed point for A.*

Proof. Assume that $Ax_0 = x_1 \neq x_0$. There exist open neighborhoods $U(x_0)$ of x_0 and $U(x_1)$ of x_1 such that $AU(x_0) \subset U(x_1)$ and $U(x_0) \cap U(x_1) = \emptyset$. Moreover, we may also take $U(x_0)$ such that some neighborhoods U_1 of $\overline{U(x_0)}$ and U_2 of $\overline{U(x_1)}$ will not intersect: $U_1 \cap U_2 = \emptyset$. Because x_0 is an essential point of φ, the set $\varphi^{-1}(U(x_0)) = M_0 \subset M$ is and essential set. Therefore, for every $\varepsilon > 0$, $g(M_0)_\varepsilon \cap (M_0)_\varepsilon \neq \emptyset$. Also $gM_o = \varphi^{-1}A(U(x_0))$ and, by uniform continuity of g, for every $\delta > 0$ one finds $\varepsilon > 0$ such that $g(M_0)_\varepsilon \subset [\varphi^{-1}AU(x_0)]_\delta$. By uniform continuity of φ, we may choose $\delta > 0$ and after $\varepsilon > 0$ such that $\varphi((M_0)_\varepsilon) \subset U_1$ and $\varphi[g(M_0)_\varepsilon] \subset U_2$ (recall that $AU(x_0) \subset U(x_1)$). Then $U_1 \cap U_2 \neq \emptyset$ which contradicts our construction. □

Note that Theorems 5.3 and 7.1 form [GM] are immediate consequences of the above theorem. (However, as was noted in section 1, this is just a different interpretation of the same approach as in [GM].)

5. Finite Dimensional Case.

Let (X, ρ, μ) be a compact metric space with a metric ρ and a probability measure μ and let $\alpha(X;\varepsilon)$ be the concentration function of X (see section 3). Let also G be a family of measure-preserving maps of X into X. (It could be that G contains one map only.)

Definition. *The pair $(X;G)$ acts on a compact metric space K via a map $\varphi : X \to K$ if for every $g \in G$ there exists $A_g : K \to K$ such that*

$$A_g\varphi = \varphi g \ .$$

(i.e., φ is an equivariant map).

Note that if a group (or semigroup) G acts on K then fixing $x^0 \in K$ and considering the map $\varphi : G \to K$, $\varphi(g) = gx^0$, we arrive to an action of the pair $(G;G)$ in the above sense.

We say that the action is $(C;C_1)$-Lipschitz iff

i) $\rho(A_g x, A_g y) \le C\rho(x,y)$ for every A_g and $x, y \in K$

and

ii) $\rho\big(\varphi(x), \varphi(y)\big) \le C_1\rho(x,y)$ for every $x, y \in K$.

Let $N_K(r)$ define the smallest number of balls of radius r which cover K and $\varepsilon_X = \inf\{\varepsilon > 0 : \alpha(X, \varepsilon/2) < 1/2\}$. Note that in Examples 3.1.i) and 3.1.ii) $\varepsilon_X = -$. However, it is not so in discrete examples such as 3.1.iii) and 3.1.iv) where $\varepsilon_X = 2\inf\{\rho(x,t),\ s \ne t \text{ from } X\} = \frac{2}{n}$ for $X = \Pi_n$ or $= E_2^n$.

Theorem 5.1. *Under the above conditions and notations there exists $x_0 \in K$ such that for every A_g, $g \in G$,*

$$\rho(A_g x_0, x_0) \le t = \inf_{\varepsilon > \varepsilon_X} \big(r(\varepsilon) + C_1\varepsilon\big)(C+1)$$

where $r(\varepsilon)$ is the smallest number satisfying an inequality

5.1
$$N_K(r) \le 1/\alpha(X;\varepsilon/2) \ .$$

(We call such x_0 a t-fixed point of the action $(\varphi : X \to K; G)$.)

Proof: Fix $\varepsilon > 0$ such that $\alpha(X;\varepsilon/2) < 1/2$ and $r > 0$ satisfying (5.1). Let $\mathcal{N}$ be a subset of K such that balls $B_r(x)$ of radius r with $x \in \mathcal{N}$ cover K and $|\mathcal{N}| = N = N_K(r)$. Choose $x_0 \in \mathcal{N}$ such that

$$\mu(\varphi^{-1}B_r(x_0)) \geq 1/N \geq \alpha(X;\varepsilon/2) .$$

We will prove that x_0 is a t-fixed point of the action $(\varphi : X \to K, G)$ for $t = (r + C_1\varepsilon)(C+1)$.

Define $M = \varphi^{-1}B_r(x_0)$. By Lemma 3.2, $\mu(M_\varepsilon) \geq 1 - \alpha(X;\varepsilon/2) > 1/2$ and $\mu(gM_\varepsilon) = \mu(M_\varepsilon) > 1/2$ for every $g \in G$. Also $\varphi(M_\varepsilon) \subset B_{r+C_1\varepsilon}(x_0)$ (by ii)) and

$$A_g B_{r+C_1\varepsilon}(x_0) \subset B_{C(r+C_1\varepsilon)}(A_g x_0)$$

(by i)). Therefore

$$\mu(\varphi^{-1}B_{r+C_1\varepsilon}(x_0)) \geq \mu(M_\varepsilon) > 1/2$$

and

$$\mu(\varphi^{-1}B_{C(r+C_1\varepsilon)}(A_g x_0)) \geq \mu(\varphi^{-1}A_g B_{r+C_1\varepsilon}(x_0)) =$$
$$= \mu(g\varphi^{-1}B_{r+C_1\varepsilon}(x_0)) > 1/2 .$$

So $B_{r+C_1\varepsilon}(x_0) \cap B_{C(r+C_1\varepsilon)})A_g x_0) \neq \emptyset$ and $\rho(x_0, A_g x_0) \leq (C+1)(r + C_1\varepsilon) = t$. Because $g \in G$ is any map form the family G, the theorem is proved.

Remark 5.2. If instead of i) and ii) we have a Hölder action of $(\varphi : X \to K; G)$, i.e.,

i' : $\rho(A_g x, A_g y) \leq C\rho(x,y)^\delta$ for some $0 < \delta \leq 1$

and any $x, y \in K$, $g \in G$,

ii' : $\rho(\varphi(x), \varphi(y)) \leq C_1\rho(x,y)^{\delta_1}$ $\quad (0 < \delta_1 \leq 1)$

for every $x, y \in X$, then there exists a t-fixed point of the action $(\varphi; G)$ for

$$t = \inf_{\varepsilon_X < \varepsilon} \left[r(\varepsilon) + C_1\varepsilon^{\delta_1} + C\big(r(\varepsilon) + C_1\varepsilon^{\delta_1}\big)^\delta\right]$$

where, as before, $N_K\big(r(\varepsilon)\big) \simeq 1/\alpha(X;\varepsilon/2)$.

Corollary 5.3. Let in the Theorem 5.1 $X = G$ be a compact group. Then the radius of a minimal invariant subset of the action G on K is at most

$$t = \inf_{\varepsilon > \varepsilon_G} \big(r(\varepsilon) + C_1\varepsilon\big)(C+1) \tag{5.2}$$

where, as before $r(\varepsilon)$ is the smallest number satifying an inequality $N_K(r) \leq 1/\alpha(G;\varepsilon/2)$.

Corollary 5.4. If in the Theorem 5.1 K is a minimal closed invariant set of the action $(\varphi : X \to K, G)$ and G is a semigroup then the radius of K is at most t as given by (5.2).

Example 5.5. Let K be a compact set in $(\mathbb{R}^k, \|\cdot\|)$ for some norm $\|\cdot\|$ and let K contain no proper closed invariant subset of a $(C;C_1)$-Lipschitz action $(\varphi : X \to K, G)$ with G being a semigroup. If $\alpha(X;\varepsilon) < a\exp(-b\varepsilon^2 n)$ then, for

$$k \geq \varepsilon_X b\, n/(4\ln[6(C+1)a^{1/k}]) \,, \tag{5.3}$$

one has an upper bound for $R = \text{Radius } K$

$$R \leq f(C, C_1, a, b)\sqrt{k/n} \tag{5.4}$$

where $f(C,C_1,a,b) = 4C_1(C+1)\sqrt{[\ln 6a(C+1)]/b}$.

Proof: Use Corollary 5.4. In our case it is well known that $N_K(r) \leq (1+2R/r)^k$ (see, e.g., [F.L.M.]). Therefore, we will satisfy (5.1) if

$$3Ra^{1/k}e^{-b\varepsilon^2 n/4k} \leq r \,.$$

so, from (5.2) it follows that

$$R \leq \inf_{\varepsilon > \varepsilon_X} (3Ra^{1/k}e^{-b\varepsilon^2 n/4k} + C_1\varepsilon)\,(C+1)\,.$$

Choose ε such that $\varepsilon^2 = \frac{4k}{bn}\ln[6(C+1)a^{1/k}]$, which implies (5.4).

Remark. Condition (5.3) often disappears because, as we noted before, $\varepsilon_X = 0$ in Examples 3.1.i) and 3.1.ii) . Also, e.g., in the interesting cases $X = E_2^n$ (see 3.1.iii) of $X = \Pi_n$ (see 3.1.iv) we have $\varepsilon_X = 2/n$ and (5.3) becomes just $k \geq 1$.

References

[AM] D. Amir, V.D. Milman. A quantitative finite-dimensional Krivine theorem. Isr. J. Math. (1985).

[FLM] T. Figiel, J. Lindenstrauss, V.D.Milman. The dimensions of almost spherical sections of convex bodies. Acta Math., 139 (1977), 53-94.

[G] M. Gromov. Filling Riemannian manifolds. J. of Differential Geometry, 18 (1983), 1-147.

[GM] M. Gromov, V.D. Milman. A topological application of the isoperimetric inequality. Am. J. Math., 105 (1983), 843-854.

[L] P. Levy. Problèmes Concrets d'Analyse Fonctionnelle. Gauthier-Villars, Paris, 1951.

[M] B. Maurey. Construction de suites symétriques. Comptes Rendus Acad. Sci. Paris, 288 (1979), A. 679-681.

[MSch] V.D. Milman, G. Schechtman. Asymptotic Theory of Finite Dimensional Normed Spaces. Springer-Verlag, Lecture Notes in Mathematics 1200, 156pp..

[Sch] G. Schechtman. Levy type inequality for a class of finite metric spaces, in Martingale Theory in Harmonic Analysis and Applications. Cleveland 1981, Springer Lecture Notes in Math. No. 939, 1982, pp. 211-215.

THE RELATION BETWEEN THE DISTANCE AND THE WEAK DISTANCE FOR SPACES WITH A SYMMETRIC BASIS

JORAM LINDENSTRAUSS and ANDRZEJ SZANKOWSKI
Department of Mathematics
The Hebrew University, Jerusalem

1. Survey

The notion of the Banach Mazur distance

$$(1) \qquad d(X,Y) = \inf \{ \|T\| \, \|T^{-1}\| \, ; T : X \to Y \}$$

between two Banach spaces X and Y with $\dim X = \dim Y = n$ is of course a basic notion in the local theory of Banach spaces. Unfortunately, it is often very hard to evaluate the quantity (1) even for concrete X and Y. The difficulty stems from the general problem of evaluating the norm of an operator and especially from the fact that it is very difficult to predict the form of the operator T for which the inf in (1) is attained or almost attained.

In order to represent an operator T from X to Y in a convenient way we have to choose convenient coordinate systems in X and Y and then give the matrix representation of T with respect to these coordinate systems. These two steps, of choosing convenient coordinates and then "good" matrices enter often also in the evaluation of $d(X,Y)$. There are several papers in the literature which deal with producing convenient coordinate systems in X and Y in connection with the computation of $d(X,Y)$ (we single out in particular the paper [B.M.]) Our purpose here is to study the problem of evaluating (1) for spaces X and Y where the choice of coordinate systems provides no problem. We shall assume that both X and Y have a 1-symmetric basis. The symmetric basis forms a natural and an essentially canonical (i.e. unique) coordinate system (cf. e.g. [R] Th. IX.8.3 and [S.1]; however, note Problem 6 below). The problem of evaluating (1) in our setting reduces just to that of choosing good matrices. As we shall see even in this case the distance is not always known. We feel that a better understanding of the problem in this rather special situation is interesting and its solution should contribute to a better understanding of the situation for general spaces.

In a recent paper we evaluated (approximately) the weak distance - to be defined below - between general spaces with a symmetric basis. In this report we shall survey the results obtained in [L.S.] and present the information we have concerning the evaluation of the distance itself. In this first section we state and explain the known results and mention some specific problems concerning the evaluation of the distance. In the second section we shall present the proofs of those of the results mentioned in section 1 which were not already proved in [L.S.].

We start with definitions which are essential for a proper formulation of the situation. Let $\{e_i\}_{i=1}^n$ be a normalized symmetric basis of X, i.e. $\|e_i\| = 1$ and

$$(2) \qquad \|\sum_{i=1}^{n} a_i e_i\| = \|\sum_{i=1}^{n} \theta_i a_i e_{\pi(i)}\|$$

for every choice of scalars $\{a_i\}_{i=1}^n$, signs $\{\theta_i\}_{i=1}^n$ and a permutation π of $\{1,\ldots,n\}$. We put

(3) $$\varphi_X(k) = \| \sum_{i=1}^k e_i \| \qquad 1 \le k \le n.$$

The function φ_X determines X up to a logarithmic factor. Indeed it is well known and easy to see that if $\varphi_X \equiv \varphi_Y$ then the identity map from X to Y has a norm at most $3\log n$ for large n (log without a specific base means log to the base 2). The verification of this remark is by decomposing general vectors into characteristic functions. Since this is used over and over again in this subject we present the trivial details.

Let $u = \sum_{i=1}^n a_i e_i$ with $\|u\|_X = 1$ and put

$$\sigma_j = \{i\,;\, 2^{-j} < |a_i| \le 2^{-j+1}\}, \qquad 1 \le j \le [\log n] + 1.$$

For every j, $2^{-j}\varphi_X(|\sigma_j|) \le 1$ where $|\sigma_j|$ denotes the cardinality of σ_j. Thus

$$\|u\|_Y \le \sum_{j=1}^{[\log n]+1} \| \sum_{i\in\sigma_j} a_i e_i \|_Y + \Sigma' |a_i|$$
$$\le \sum_{j=1}^{[\log n]+1} 2^{-j+1}\varphi_Y(|\sigma_j|) + n\cdot\frac{1}{n} \le 3 + 2[\log n]$$

where in Σ' above we sum over those indices i for which $|a_i| \le 2^{-[\log n]-1}$.

We extend the definition of φ_X to a function defined on $[1,n]$ by making it linear on each interval $[k,k+1]$. As in [G.1] we define a function f_X on $[0,1]$ by putting

(4) $$f_X(t) = \log_n \varphi_X(n^t) - t/2$$

where $\log_n$ means, of course, log to the base n. The reason for introducing f_X is that it will enable the statement of results and problems on n-dimensional symmetric spaces in a way where n does not appear explicitly in the formulas. It follows easily from (2) and the triangle inequality that f_X satisfies

$$f_X(0) = 0, \quad |f_X(s) - f_X(t)| \le \frac{1}{2}|s-t|, \quad 0 \le s,t \le 1.$$

Conversely, it is easy to see (cf. [L.S.]) that for every function f on $[0,1]$ satisfying

(5) $$f(0) = 0, \quad |f(s) - f(t)| \le \frac{1}{2}|s-t|, \quad 0 \le s,t \le 1$$

there is for every integer n an n-dimensional space X with a symmetric basis so that $f(t) = f_X(t)$ for every $t \in [0,1]$ such that n^t is an integer. In other words there is for every integer n a correspondence between the class of n-dimensional spaces with a symmetric basis and the class $\mathcal{F}$ of functions satisfying (5). This correspondence is one to one up to logarithmic factors, since X is determined by f, in the Banach Mazur distance, up to $c\,\log^2 n$ while

$f \in \mathcal{F}$ is determined by X at the points $\{\log k/\log n\}_{k=1}^{n}$ and thus at every point t up to an additive term of size $\leq c/\log n$.

The function corresponding to ℓ_p^n, $1 \leq p \leq \infty$, is $t(\frac{1}{p} - \frac{1}{2})$ i.e. a linear function. In particular to the Hilbert space ℓ_2^n corresponds the function which is identically 0. The slope of f_X is closely connected to the type or cotype of X. For example it is easy to check that if the cotype 2 constant of X is bounded by c then there is a monotone increasing function f on [0,1] so that $|f(t)-f_X(t)| \leq 2\log c/\log n$ for all t, and similarly if X has type 2, then f_X is "almost" decreasing. Another trivial remark concerning the correspondence $X \longleftrightarrow f$, which follows immediately from the definitions, is that for all X

(6) $\qquad f_X(t) = -f_{X^*}(t), \quad 0 \leq t \leq 1.$

The weak distance between two Banach spaces X and Y was defined in [T.J.2] as follows: Put

(7) $$q(X,Y) = \inf\{\int_\Omega \|T(\omega)\| \, \|S(\omega)\| d\omega\}$$

where the inf is taken over all measure spaces Ω, all (measurable, of course) maps T (resp. S) from Ω to the space of bounded linear operator from X to Y (resp. from Y to X) such that

$$\int_\Omega S(\omega)T(\omega)d\omega = I_X = \text{identity of } X.$$

The weak distance $\hat{d}(X,Y)$ is defined to be $q(X,Y) \vee q(Y,X)$ (we use $\vee$ and $\wedge$ for the usual lattice operations). The notion of weak distance is closely connected to the theory of operator ideals but we shall not consider this aspect here. We just mention that, roughly speaking, $\hat{d}(X,Y)$ measures how far the "natural" parameters of X can be from those of Y.

It was observed in [L.S.] that if X and Y are n-dimensional spaces with a symmetric basis (actually all what is needed is that they have enough symmetries in the sense of [G.G.]) then

(8) $$q(X,Y) = \hat{d}(X,Y) = \inf\{\|T\| \, \|S\| \,;\, T: X \to Y,\, S: Y \to X,\, \text{trace } ST = n\}.$$

Most of the estimates for $\hat{d}(X,Y)$ or $d(X,Y)$ for spaces with a symmetric basis are of the form $n^{\eta(f_X,f_Y)}$ up to logarithmic factors where $\eta(f,g)$ is an expression depending only on the functions f and g. Therefore in order to get results which are independent of n in their formulation it is natural to consider $\log_n \hat{d}(X,Y)$ and $\log_n d(X,Y)$. A logarithmic multiplicative term in $d(X,Y)$ or $\hat{d}(X,Y)$ becomes after passing to $\log_n$ an additive term of the form $O(\log\log n/\log n)$. By this we mean a term of absolute value less than $c \log\log n/\log n$ where c is an absolute constant (independent not only of n but also of f_X and f_Y).

The main result of [L.S.] is the following.

Theorem: Let X and Y be n-dimensional spaces with a symmetric basis. Then

(9) $$\log_n \hat{d}(X,Y) = \min\{M(f_X, f_Y, z, s, u) + M(-f_X, -f_Y, z, s, u);$$

$$0 \le s \le 1,\ 0 \le u \le 1,\ 0 \le z \le s \wedge u\} + O(\log\log n / \log n)$$

where

(10) $$M(f, g, z, s, u) = [\max_{0 \le w \le z} \{g(w) - f(w+u-z)\} + (1-s)/2] \vee$$

$$\vee [\max_{z \le w \le s \wedge u} \{g(w) - f(u)\} + (1-s)/2] \vee$$

$$\vee [\max_{0 \le v \le z} \{g(v+s-z) - f(v)\} + (1-u)/2] \vee$$

$$\vee [\max_{z \le v \le s \wedge u} \{g(s) - f(v)\} + (1-u)/2].$$

The formulas are obviously complicated. It is much easier to understand the operators which are used in obtaining them. The proof of the observation (8) also shows that for spaces with a symmetric basis

(11) $$\hat{d}(X,Y) = \inf\{\int_\Omega \|T(\omega)\|\ \|S(\omega)\| d\omega ; \int_\Omega \operatorname{trace} S(\omega)T(\omega) d\omega = n\}.$$

The operators which were shown in [L.S.] to give the inf in (11) (up to a logarithmic factor) are of the following form: The operators depend on three parameters s, u and z and on the random variables η and ω. For $1 \le i \le n^s$, $1 \le j \le n^u$ the (i,j)'th entry of the matrix representation of the operator (denoted by $T(\eta,\omega)$ and depending also on parameters z, s and u) is equal to $d\beta_{i,j}(\eta)\gamma_{i,j}(\omega)$ where the $\beta_{i,j}$ are Bernoulli variables taking the values 1 and 0 with $\mathrm{Prob}\{\beta_{i,j} = 1\} = n^{-z}$, the $\gamma_{i,j}$ are normalized Gaussian variables, and all the $\beta_{i,j}$ and $\gamma_{i,j}$ are mutually independent. The matrix defining $S(\eta,\omega)$ is the adjoint of that defining $T(\eta,\omega)$. The constant $d = d(z,s,u)$ is chosen so that

$$\operatorname{trace} \iint S(\eta,\omega)T(\eta,\omega) d\eta d\omega = n.$$

In other words, the support of the matrix $T(\eta,\omega)$ is chosen randomly in the upper left corner of size $n^s \times n^u$ so that the expected density of the chosen indices in this rectangle in n^{-z}. On each point chosen to belong to the support the entry itself is a Gaussian variable.

The proof of the $\le$ part of (9) is carried out by analyzing the form of the support of $T(\eta,\omega)$ and using a variant of Chevet's theorem [C] which is applicable to operators with a random support. The $\ge$ part of (9) is proved by applying a combinatorial argument which shows that whenever trace $ST = n$, the matrices representing αT and $\alpha^{-1}S$ must contain in them, for a suitable α, submatrices of a sufficiently large size with many relatively large entries. By using Khintchine's inequality it is deduced that $\|T\|\ \|S\|$ must be at least as large (up to a logarithmic factor) as $\|T(\eta,\omega)\|\ |S(\eta,\omega)\|$ for a suitable choice of z, s and u and a large set of points (η,ω) in the measure space. In all the arguments the following version of the remark on reduction to characteristic functions is often used:

If $(a_{i,j})_{i,j=1}^n$ is a matrix representation of an operator T from X to Y then, up to a logarithmic factor, $\|T\|_{X\to Y}$ is given by $\sup |\sum_{i\in A}\sum_{j\in B} a_{i,j}| / \varphi_X(|A|)\varphi_{Y^*}(|B|)$ where the supremum is taken over all subsets A and B of $\{1, \ldots, n\}$.

There is an addendum to the theorem which is also proved in [L.S.] It states that the minimum in (9) is attained if either $u = 1$ or $s = 1$. In other words in the definition of $T(\eta,\omega)$ the set of size $n^s \times n^u$ in which we randomly choose the support consists either of all the rows or all the columns. The addendum is proved by analyzing the expressions (9) and (10) (without going back to the operators). To state the addendum explicitly define

(12) $$\theta(f,g) = \min\{M(f,g,z,1,u) + M(-f,-g,z,1,u);\ \ 0 \le z \le u \le 1\}.$$

The addendum states in this notation that

(13) $$\log_n \hat{d}(X,Y) = \theta(f_X, f_Y) \wedge \theta(f_Y, f_X) + O(\log\log n/\log n).$$

We pass now to $d(X,Y)$. Using matrices of the form

(14) $$\begin{bmatrix} W_1 & 0 & \cdot\ \cdot \\ 0 & W_2 & \cdot\ \cdot \\ & \cdot\ \cdot & \cdot\ \cdot \\ & \cdot\ \cdot & \cdot\ W_k \end{bmatrix}$$

where the $\{W_i\}_{i=1}^k$ are Walsh matrices of size n/k for a suitable k (n and k can, without loss of generality, be taken to be powers of 2), the following upper estimate on $d(X,Y)$ was obtained independently in [G.1] and [T.J.1].

(15) $$\log_n d(X,Y) \le \min_{0 \le t \le 1} \operatorname*{Osc}_{\Sigma_t} \{f_Y(v) - f_X(u)\} + O(\log\log n/\log n)$$

where $\operatorname*{Osc}_K h$ means $\sup\{h(\omega)\,;\,\omega \in K\} - \inf\{h(\omega)\,;\,\omega \in K\}$ and Σ_t is the subset of the unit square in the (u,v) plane marked by lines in the picture:

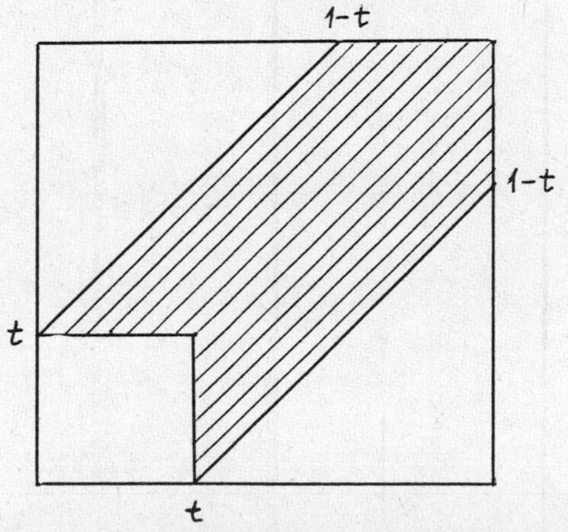

If both f_X and f_Y are monotone functions in the same sense and also if f_X and f_Y are monotone in different senses or more generally if

$$0 = f_X(0) \le f_X(u) \le f_X(1), \quad 0 = f_Y(0) \ge f_Y(u) \ge f_Y(1), \quad 0 \le u \le 1$$

then it is easy to check that the right hand sides of (13) and (15) coincide (cf. [L.S.]) and in particular $d(X,Y) = \hat{d}(X,Y)$ up to a possible logarithmic factor (the fact that in this case (15) becomes an equality was proved in [G.2] using volume considerations). Another case when this happens is the case when $f_X(t) = \pm t/2$ (i.e $X = \ell_\infty^n$ or ℓ_1^n) and f_Y is arbitrary.

What happens in other cases? It was proved in [L.S.] that

(16) $\quad \log_n d(X,Y) \ge \theta(f_X,f_Y) \vee \theta(f_Y,f_X) + O(\log\log n/\log n)$

Hence if $\theta(f_X,f_Y) \ne \theta(f_Y,f_X)$ then by (13) and (16), $d(X,Y)$ is larger than $\hat{d}(X,Y)$ by a factor of size n^δ for some $\delta > 0$. Examples of functions f and g satisfying (5) and $\theta(f,g) \ne \theta(g,f)$ exist but are not trivial to construct by hand. In [L.S.] an example of such a pair of functions was constructed. It is possible to have such a pair in which one of the functions is monotone. For example this happens if f and g are

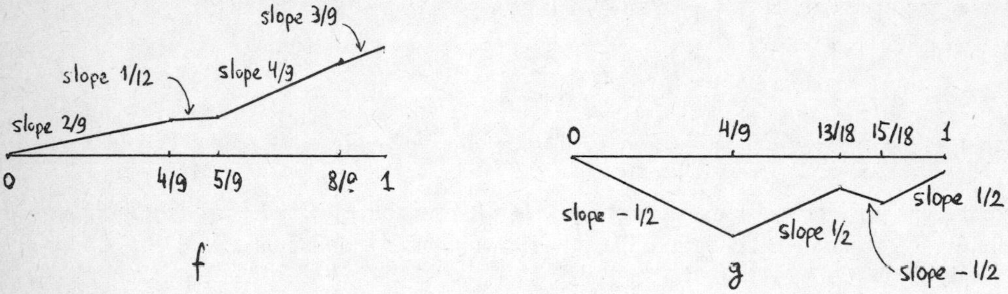

In this case $\theta(f,g) \approx 0.356$ and $\theta(g,f) \approx 0.336$.
(This example was found by using a personal computer; we are grateful to Elon Lindenstrauss for this example and related experimental work). On the other hand if f is linear and g is a general function satisfying (5), then $\theta(f,g) = \theta(g,f)$. This fact, which will be proved in the next section, is relevant to the following problem.

Problem 1. Assume that $X = \ell_p^n$ and Y a general n-dimensional space with a symmetric basis. Is it true that (even only up to a logarithmic factor) $d(X,Y)$ is equal to $\hat{d}(X,Y)$?

The results proved in the next section indicate that the answer should be positive. The answer is known to be so if $p = 1,\infty$ (as mentioned above) and also for $p = 2$ where actually it follows easily from the definition that $d(\ell_2^n,Y) = \hat{d}(\ell_2^n,Y)$ for every n-dimensional space Y (cf. [T.J.2]). Problem 1 was raised in [T.J.2] for general n-dimensional spaces Y. It is more interesting in the general context since it was proved in [T.J.2] that $\hat{d}(\ell_p^n,Y) \le c\sqrt{n}$ for some absolute constant c and every $1 \le p \le \infty$ and n-dimensional Y. In the general context the problem is open also for $p = 1,\infty$ and the answer may well be different from the special case where Y has a symmetric basis.

The proof of (16) in [L.S.] which is very close to the $\geq$ part of (9) actually shows somewhat more. In order to state it, let us introduce another quantity $\bar{d}(X,Y)$ which is by definition between $d(X,Y)$ and $\hat{d}(X,Y)$ whenever X and Y have a symmetric basis:

$$\text{(17)} \qquad \bar{d}(X,Y) = \inf\{\,\|T\|\,\|S\|;\ T: X \to Y,\ S: Y \to X,\ \operatorname{diag} ST = \operatorname{diag} TS = (1,1,\ldots,1)\}.$$

The proof of (16) actually shows that

$$\text{(18)} \qquad \log_n \bar{d}(X,Y) \geq \theta(f_X, f_Y) \vee \theta(f_Y, f_X) + O(\log\log n/\log n).$$

On the other hand the proof of the $\leq$ part of (9) uses for $s = u = 1$ (i.e. the case where the support of $T(\eta,\omega)$ is chosen randomly in the entire available space of $n \times n$ entries) operators, for which generically the diagonals of $S(\eta,\omega)\,T(\eta,\omega)$ and $T(\eta,\omega)S(\eta,\omega)$ are close to $(1,1,\ldots,1)$. Hence it follows from this argument that

$$\text{(19)} \qquad \log_n \bar{d}(X,Y) \leq \min\{M(f_X, f_Y, z, 1, 1) + M(-f_X, -f_Y, z, 1, 1);\ 0 \leq z \leq 1\} + O(\log\log n/\log n).$$

Observe, that by definition, for every pair of functions f and g we have that

$$M(f,g,z,1,1) + M(-f,-g,z,1,1) = \operatorname*{Osc}_{\sigma_{1-z}} \{g(v) - f(u)\}$$

where σ_t is the piecewise linear curve in the u,v plane given by

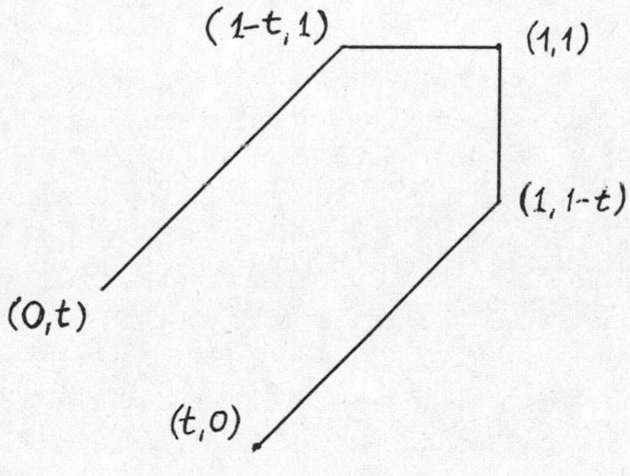

Note that σ_t is a part of the boundary of the set Σ_t defined above. Put

$$\text{(20)} \qquad \psi(f,g) = \min_{0 \leq t \leq 1} \operatorname*{Osc}_{\sigma_t}\{g(v) - f(u)\}.$$

Obviously, by definition, $\psi(f,g) = \psi(g,f)$. We can now restate (19) as

$$\text{(21)} \qquad \log_n \bar{d}(X,Y) \leq \psi(f_X, f_Y) + O(\log\log n/\log n).$$

There are examples of pairs of functions f and g satisfying (5) so that $\psi(f,g)$ is strictly larger than $\theta(f,g) \vee \theta(g,f)$ (such examples will be presented in the next section) and thus (18) and (21) together do not determine $\tilde{d}(X,Y)$ in all cases. In particular, we do not know the answer to the following two (obviously related) problems.

Problem 2. Is it true that for every n-dimensional spaces X and Y with a symmetric basis

$$\log_n \tilde{d}(X,Y) = \psi(f_X, f_Y) + O(\log\log n/\log n)?$$

Problem 3. Is it true that for every n-dimensional spaces X and Y with a symmetric basis

$$\log_n d(X,Y) \geq \psi(f_X, f_Y) + O(\log\log n/\log n)?$$

Another open problem concerning the distance between spaces with a symmetric basis, which is very natural but seems to be open, is

Problem 4. Let X and Y be n-dimensional spaces with a symmetric basis. Is the distance between X and Y achieved (at least up to a logarithmic factor) by considering only orthogonal operators T from X onto Y?

In all the examples we checked the answer is yes. It is easy to see (and we shall prove this in the next section) that a positive answer to problem 4 implies a positive answer to problem 3. However, a positive answer to problem 4 by itself will be only of a limited value since it is a priori not clear what orthogonal matrices should be used for computing $d(X,Y)$.

We mentioned the matrices of type (14) which led to formula (15) and thus to the computation of $d(X,Y)$ in many cases. The conjecture that these matrices will always suffice for determining $d(X,Y)$, at least up to logarithmic factors (cf. [S.2]), is however false. By using orthogonal matrices which do not contain in them submatrices which are "almost" as big (as far as the size of the support and the size of entries in the support are concerned) as a Walsh matrix of any size, it is possible to construct n dimensional spaces with symmetric bases X and Y so that $\log_n d(X,Y)$ is strictly smaller than the right hand side of (15). Bela Bollobas turned our attention to one such family of matrices, namely the orthogonal matrices which correspond to the incidence matrices of finite projective planes. We are grateful to him for his suggestion, which enabled us to obtain the following new estimate for $d(X,Y)$ (see the next section for a proof).

(22)
$$\log_n d(X,Y) \leq \max_{(u,v)\in\Pi}\{f_Y(v)-f_X(u)\} \vee \max_{0\leq u\leq 1/2}\{f_Y(u)-f_X(u)-\frac{1}{4}+\frac{u}{2}\} + $$
$$+ \max_{(u,v)\in\Pi}\{f_X(v)-f_Y(u)\} \vee \max_{0\leq u\leq 1/2}\{f_X(u)-f_Y(u)-\frac{1}{4}+\frac{u}{2}\}$$
$$+ O(\log\log n/\log n)$$

where Π is the set in the (u,v) plane, given by

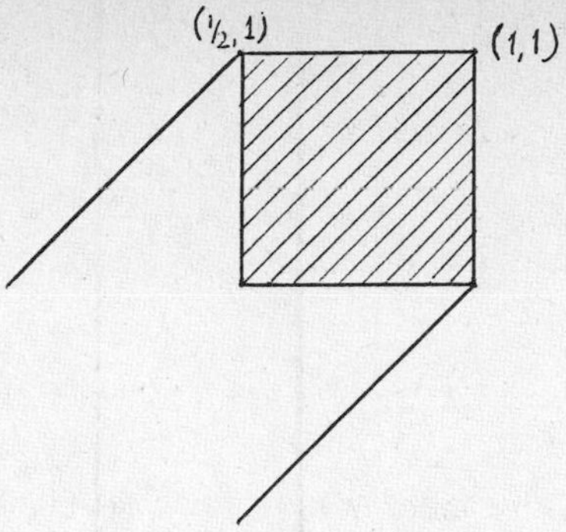

(it is the union of a square with two segments). Notice that

$$\sigma_{1/2} \subset \Pi \subset \Sigma_{1/2}.$$

If $f_X(u) \le f_X(\tfrac12)$ and $f_Y(u) \le f_Y(\tfrac12)$ for $0 \le u \le \tfrac12$, then, by (5), the right hand side of (22) reduces to $\operatorname{Osc}_{\Pi}\{f_Y(v) - f_X(u)\}$.

For suitable f_X and f_Y the right hand side of (22) is strictly smaller than the right hand side of (15). By using other families of matrices we get additional upper estimates for $d(X,Y)$. However at present we know only examples and by using our examples we are unable to push the estimate of $\log_n d(X,Y)$ all the way down to $\psi(f_X, f_Y)$. Therefore we state:

Problem 5. Are there examples of n-dimensional spaces X and Y with a symmetric basis so that

$$\log_n d(X,Y) > \psi(f_X, f_Y) + O(\log\log n / \log n)?$$

In view of (21) what is needed for finding such X and Y (if indeed they exist) is a lower estimate on $d(X,Y)$ which uses the fact that in the matrix TT^{-1} the off diagonal elements are 0. All we used till now is that the diagonal elements of TT^{-1} are all 1 and this only estimates $\bar{d}(X,Y)$.

Problem 5 is related to the question of uniqueness of symmetric bases which is, in spite of much available information (cf. [J.M.S.T.] and [S.1]) not completely solved. We state here a version of this problem in the setting of our discussion (i.e. equality up to logarithmic factors).

Problem 6. Given constants C and α do there exist D and β so that whenever X and Y are n-dimensional spaces with a symmetric basis for which $d(X,Y) \le C(\log n)^{\alpha}$, then the identity map $I_{X\to Y}$ from X to Y has norm at most $D(\log n)^{\beta}$?

As pointed out in [L.S.] what we can deduce from the fact that $d(X,Y) \le C(\log n)^{\alpha}$ is that

(23) $$\| I_{X\to Y} \| \le e^{\gamma(\log\log n \cdot \log n)^{1/2}}$$

for some constant $\gamma=\gamma(C,\alpha)$. As also mentioned in [L.S.] it was pointed out to us by J. Bourgain that if the answer to Problem 5 is negative, then (23) cannot be improved at all and in particular also the answer to Problem 6 is then negative. It should be mentioned that Problem 6 is open only for pairs X,Y which are "very near" to ℓ_2^n (cf [S.1] and [L.S.])

2. Some proofs

We start this section with a proof of (22). Let p be a power of a prime and consider the projective plane over the field having p elements. The plane has $n=p^2+p+1$ points and lines. Let $\{\alpha_{i,j}\}_{i,j=1}^n$ be the incidence matrix of this plane (i.e. $\alpha_{i,j}=1$ if point i is on line j and $\alpha_{i,j}=0$ otherwise). Let T be the $n\times n$ matrix whose entries $a_{i,j}$ are given by

$$(24)\qquad a_{i,j}=\begin{cases}(\sqrt{p}+1)/(1+\sqrt{p}+p) & \text{if} \quad \alpha_{i,j}=1\\ -1/\sqrt{p}\,(1+\sqrt{p}+p) & \text{if} \quad \alpha_{i,j}=0\end{cases}$$

It is trivial to check that (24) defines an orthogonal matrix. Such a matrix has already been used in Banach space theory before (see e.g. [K]). In order to estimate the norm of T and $T^*=T^{-1}$ as an operator between symmetric spaces we need the following, certainly known, combinatorial lemma

Lemma. Let $1\le h\le k\le n$ and put

$$(25)\qquad \tau(h,k)=\max\{\sum_{i\in A}\sum_{j\in B}\alpha_{i,j}\ ;\ |A|=h,\ |B|=k\},$$

i.e. the maximal number of incidences between h points and k lines. Then, for some absolute constant c

$$(26)\qquad \tau(h,k)\le c\,([hk^{1/2}\vee k]\wedge hn^{1/2}).$$

Proof. Since every line contains $p+1$ points we clearly have that $\tau(h,k)\le 2hn^{1/2}$ so that one half of (26) is obvious. To prove the second half put $\tau_r=\tau(2^r,2^{2r})$ for $r=1,2,\cdots$. Denote by Δ the support of the matrix $\{\alpha_{i,j}\}_{i,j=1}^n$. Let A and B be sets with $|A|=2^r$, $|B|=2^{2r}$ so that $\tau_r=|(A\times B)\cap\Delta|$. We partition A into $A'\cup A''$ with $|A'|=|A''|=2^{r-1}$. Let $B',B''\subset B$ be such that for every $j\in B'$ (resp. every $j\in B''$) there are at least two points i in A' (resp. A'') such that $(i,j)\in\Delta$. Since through every two different points passes exactly one line there are at most $\binom{2^{r-1}}{2}$ elements in B' or B''. Since $\binom{2^{r-1}}{2}\le 2^{2(r-1)}$, we have

$$\tau_r\le|(A'\times B')\cap\Delta|+|(A''\times B'')\cap\Delta|+2\cdot 2^{2r}\le 2\tau_{r-1}+2\cdot 2^{2r},$$

which implies, by induction, that $\tau_r\le 4\cdot 2^{2r}$. This proves (26) if $h=2^r$ and $k=2^{2r}$ for some r and hence, by changing c somewhat, whenever $k=h^2$.

Assume now that $k > h^2$. Then by singling out those lines which contain more than one point in a given set A with $|A| = h$ we get that

$$\tau(h,k) \le \tau(h,h^2) + k - h^2 \le ch^2 + k - h^2 \le ck.$$

Finally, if $h \le k \le h^2$ put $h' = [k^{½}] + 1$. We have

$$\tau(h,k) \le \left(\frac{h}{h'} + 1\right)\tau(h',k) \le c_1 hk^{½}$$

for a suitable constant c_1.

□

Proposition 1. Let T be the operator whose matrix representation is given by (24). Let X and Y be n-dimensional spaces with a symmetric basis. Then

$$(27) \qquad \log_n \|T\|_{X\to Y} \le \max_{(u,v)\in\Pi} (f_Y(v) - f_X(u)) \vee \max_{0\le u\le ½} \left(f_Y(u) - f_X(u) - \frac{1}{4} + \frac{u}{2}\right) + O(\log\log n/\log n)$$

Proof. Let $h \le k$ and assume also that $h \le n^{½}$. Since by (24), $|a_{i,j}| \sim n^{-1/4}$ if $(i,j) \in \Delta$ and $|a_{i,j}| \sim n^{-3/4}$ if $(i,j) \notin \Delta$, we get from (26) that whenever $|A| = h$, $|B| = k$

$$(28) \qquad \left|\sum_{i\in A}\sum_{j\in B} a_{i,j}\right| \le \sum_{i\in A}\sum_{j\in B} |a_{i,j}| \le \begin{cases} chk^{½}n^{-1/4} & \text{if } h \le k \le h^2 \\ ckn^{-1/4} & \text{if } h^2 \le k \le n^{½}h \\ chn^{1/4} & \text{if } n^{½}h \le k \end{cases}$$

where c is a universal constant. If $n^{½} \le h \le k$ we get by the orthogonality of $\{a_{i,j}\}$ that

$$(29) \qquad \left|\sum_{i\in A}\sum_{j\in B} a_{i,j}\right| \le (hk)^{½}.$$

If $k < h$ we get estimates similar to (28) and (29) provided we exchange the role of h and k. As mentioned in section 1, $\|T\|_{X\to Y}$ is, up to a logarithmic factor, equal to

$$(30) \qquad \max_{1\le h,k\le n} \max\left\{\left|\sum_{i\in A}\sum_{j\in B} a_{i,j}\right| / \varphi_X(h)\varphi_{Y^*}(k)\,;\ |A| = h,\ |B| = k\right\}.$$

Since $h/\varphi_X(h)$ and $k/\varphi_{Y^*}(k)$ are increasing functions of h respectively k, it follows that if we use (28) in order to estimate (30) for $h \le k \wedge n^{½}$, we get the largest expression if $h = k$ or $k = n^{½}h$. Similarly, if $k \le h \wedge n^{½}$, we get the largest expression if $h = k$ or $h = n^{½}k$. If $h \wedge k > n^{½}$ we estimate (30) by using (29). Putting all these facts together and passing to $\log_n$ we get exactly the desired result i.e (27).

□

The proof of Proposition 1 is quite simple since in the crucial case $h < n^{½}$ we have practically no cancellations in $\sum_{i\in A}\sum_{j\in B} a_{i,j}$ and we can estimate this expression in (28) by passing to $|a_{i,j}|$. For most operators T the cancellations play a crucial role and the estimates are generally more difficult.

Since $T^* = T^{-1}$ for T given by (24), we get that the estimate for $\|T^{-1}\|_{Y \to X}$ is the same as (27); we have only to replace f_Y by f_X and vice versa. Hence (22) is an immediate consequence of Proposition 1. An example where (22) gives a better estimate than (15) is obtained by taking

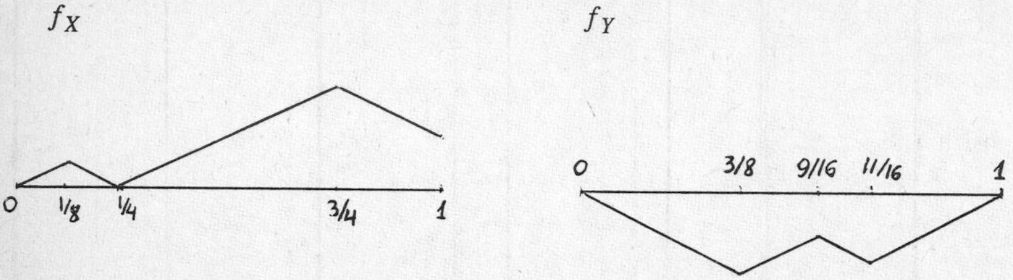

(in both functions all the slopes are $\pm½$). For these functions the right hand side of (15) is 10/32 while that of (22) is 9/32 (plus, of course, $O(\log\log n/\log n)$).

A direct inspection of (22) shows that if one of the functions, say f_X, is monotone on [0,1], then the right hand side of (22) becomes $\operatorname*{Osc}_{\sigma_{½}}\{f_Y(v) - f_X(u)\}$, (ignoring again the $O(\log\log n/\log n)$ term). By using the lemma, and $n \times n$ matrices of the form

$$(31) \qquad T_m = \begin{bmatrix} T & 0 & & \\ 0 & T & & \\ & & & 0 \\ & & 0 & T \end{bmatrix}$$

where T is an $m \times m$ orthogonal matrix corresponding to the incidence matrix of a suitable projective plane, we can get, like in Proposition 1, upper estimates for $d(X,Y)$. If $m = n^s$, $0 < s < 1$, and if in particular f_X is monotone on [0,1] we get as an upper estimate for $\log_n d(X,Y)$ the expression $\operatorname*{Osc}_{\sigma_{s/2}}\{f_Y(v) - f_X(u)\}$. Consequently, if either f_Y or f_X is monotone on [0,1], then we have

$$(32) \qquad \log_n d(X,Y) \le \min_{0 \le t \le ½} \operatorname*{Osc}_{\sigma_t}\{f_Y(v) - f_X(u)\} + O(\log\log n/\log n).$$

We have examined also other examples of orthogonal matrices (e.g. the tensor product of (24) with Walsh matrices, or matrices corresponding to incidence matrices of points and hyperplanes in projective spaces of dimension greater than 2) however we do not have a matrix which will show that if, say, f_X is monotone, then $\log_n d(X,Y)$ is bounded from above by $\operatorname*{Osc}_{\sigma_t}\{f_Y(v) - f_X(u)\}$ also for $½ < t < 1$. This is exactly the missing link in our attempt to solve problem 1.

The next proposition, whose proof involves just an examination of formula (10) and no direct Banach space arguments, is also obviously related to problem 1.

Proposition 2. Let f and g be functions satisfying (5) with g being linear. Then

$$\theta(f,g) = \theta(g,f) = \min_{0\le t\le 1} \operatorname*{Osc}_{\sigma_t}\{g(v)-f(u)\} = \psi(f,g).$$

Proof. We assume that $g(t) = at$, $0 \le t \le 1$, and we also assume without loss of generality that $0 \le a \le ½$. We show first that $\theta(g,f) = \psi(f,g)$. In view of the definitions of θ and ψ this will follow once we prove that whenever $0 \le z \le s < 1$, then

$$M(f,g,z,s,1) + M(-f,-g,z,s,1) \ge M(f,g,z,1,1) + M(-f,-g,z,1,1).$$

Put

(33) $\qquad F(x) = M(f,g,z,x,1) + M(-f,-g,z,x,1)$ for $s \le x \le 1$.

In order to verify that $F(s) \ge F(1)$ it suffices to show that $\partial F(x) \le 0$ for $x \in [s,1]$ where we define for any function h

$$\partial h(x) = \lim_{\varepsilon\to 0} \sup\, (h(x+\varepsilon)-h(x))/\varepsilon.$$

From the definition of F it is clear that it suffices to prove that $\partial F(s) \le 0$.

We define now some subsets of the unit square

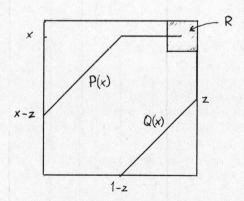

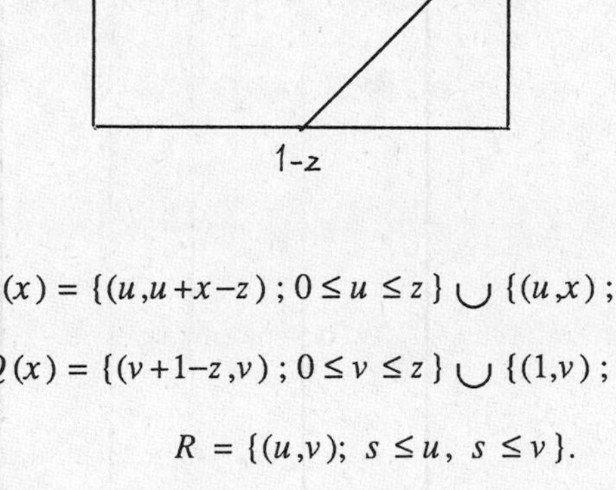

$$P(x) = \{(u,u+x-z)\,;\, 0 \le u \le z\} \cup \{(u,x)\,;\, z \le u \le x\}$$

$$Q(x) = \{(v+1-z,v)\,;\, 0 \le v \le z\} \cup \{(1,v)\,;\, z \le v \le x\}$$

$$R = \{(u,v);\ s \le u,\ s \le v\}.$$

Next, put

$$\lambda^{\pm}(x) = \max\{\pm(g(v)-f(u));\quad (u,v) \in P(x)\}$$

$$\mu^{\pm}(x) = \max\{\pm(g(v)-f(u)) + (1-x)/2;\quad (u,v) \in Q(x)\}$$

$$F^{+}(x) = \lambda^{+}(x) \vee \mu^{+}(x),\quad F^{-}(x) = \lambda^{-}(x) \vee \mu^{-}(x).$$

Notice that the notations were chosen so that $F(x) = F^{+}(x) + F^{-}(x)$. We claim that

(*) There is a $\delta > 0$ so that, putting

$$\tilde{\lambda}^{\pm}(x) = \max\{\pm(g(v)-f(u)); (u,v) \in P(x) \sim R\},$$

we have $F^{\pm}(x) = \tilde{\lambda}^{\pm}(x) \vee \mu^{\pm}(x)$ for $s \le x \le s + \delta$.

Indeed, it suffices to consider $F^{+}(x)$. If $\lambda^{+}(s) > g(s) - f(s)$ then by continuity, $\lambda^{+}(x) = \tilde{\lambda}^{+}(x)$ for x near s and (*) is clear. If $\lambda^{+}(s) = g(s) - f(s)$ then, since f satisfies (5)

$$\lambda^{+}(s) = g(s) - f(s) \le g(s) - f(1) + (1-s)/2 \le \mu^{+}(s).$$

If $\lambda^{+}(s) < \mu^{+}(s)$ then (*) is again clear by continuity. If $\lambda^{+}(s) = \mu^{+}(s)$ then the inequality above implies that $f(t) = f(s) + (t-s)/2$ for $s \le t \le 1$. In particular f is increasing on $[s,1]$ and hence $\tilde{\lambda}^{+}(x) = \lambda^{+}(x)$ for all $x \ge s$, and again (*) is clear.

Since $g(t) = at$ it follows that $\tilde{\lambda}^{\pm}(x) = \pm a(x-s) + \tilde{\lambda}^{\pm}(s)$ for all $x \ge s$. It follows also easily from the definition of μ that $\partial\mu^{+}(s) \le a - ½$ while $\partial\mu^{-}(s) = -½$. Hence, by (*),

$$\partial F^{+}(s) \le a \vee (a - ½) = a,\quad \partial F^{-}(s) \le -a \vee -½ = -a$$

and $\partial F(s) = \partial F^{+}(s) + \partial F^{-}(s) \le a - a = 0$ as desired.

We turn to the second part of the proposition i.e. to the proof that $\theta(f,g) = \psi(f,g)$. To this end it suffices to prove that whenever $0 \le z < s < 1$, then $G(s) \ge G(1)$ where

(34) $G(x) = M(f,g,z+x-s,1,x) + M(-f,-g,z+x-s,1,x)$ for $s \le x \le 1$

(Notice that in contrast to (33) here z is also changed). As before it suffices to prove that $\partial G(s) \le 0$. To prove this we introduce again some sets and functions

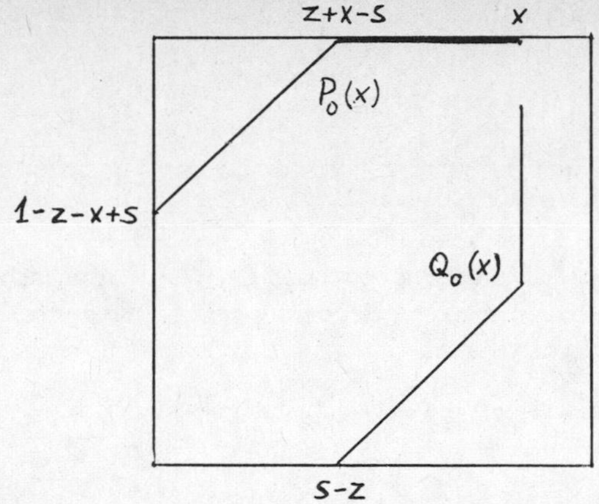

$$P_0(x) = \{(u, u+1-z-x+s);\ 0 \le u \le z+x-s\} \cup \{(u,1); z+x-s \le u \le x\}$$

$$Q_0(x) = \{(v+s-z, v);\ 0 \le v \le z+x-s\} \cup \{(x,v); z+x-s \le v \le x\}$$

$$\lambda_0^{\pm}(x) = \max\{\pm(g(v)-f(u)) + (1-x)/2;\ (u,v) \in P_0(x)\}$$

$$\mu_0^{\pm}(x) = \max\{\pm(g(v)-f(u));\ (u,v) \in Q_0(x)\}.$$

$$G^{\pm}(x) = \lambda_0^{\pm}(x) \vee \mu_0^{\pm}(x).$$

We have that $G(x) = G^+(x) + G^-(x)$. Since $g(t) = at$ is monotone increasing, it is evident that $G^+(x) = \lambda_0^+(x)$ and that

$$\mu_0^-(x) = \max\{-av + f(v+s-z);\ 0 \le v \le z+x-s\}$$

A direct inspection shows, assuming as we may that f is differentiable, that

$$\partial\mu_0^-(s) \le 0 \vee (\partial f(s) - a)$$

$$\partial\lambda_0^+(s) = [(-\partial f(s)) \vee 0] - \frac{1}{2},\quad \partial\lambda_0^-(s) = [\partial f(s) \vee a] - \frac{1}{2}.$$

Hence if $\partial f(s) > 0$, then $\partial G^+(s) = \partial\lambda_0^+(s) = -½$ and since $\partial G^-(s) \le ½$ it follows that $\partial G(s) \le 0$. If $\partial f(s) < 0$ then $\partial G^+(s) = -\partial f(s) - ½$ and $\partial G^-(s) \le 0$ and thus again $\partial G(s) \le 0$ as desired.

□

Let us remark that Propositions 1 and 2 do not suffice by themselves to settle Problem 1: It is not difficult to construct two functions, f and g which satisfy (5), g being linear so that

$$\min_{0 \le t \le ½} \underset{\sigma_t}{\mathrm{Osc}}[g(v) - f(u)] > \min_{½ \le t \le 1} \underset{\Sigma_t}{\mathrm{Osc}}[g(v) - f(u)] >$$

$$> \min_{½ \le t \le 1} \underset{\sigma_t}{\mathrm{Osc}}[g(v) - f(u)]$$

This gives a space X with a symmetric basis and a number p so that the lower bound for $d(X,\ell_p^n)$, obtained by using the matrices of the form (14) or (31) is larger than $n^\delta \hat{d}(X,\ell_p^n)$ for some $\delta > 0$.

Our next proposition is a simple remark concerning a lower bound for $d(X,Y)$ which is related to problem 4.

Proposition 3. Let T be an invertible operator from an n-dimensional space with a symmetric basis X into another such space Y. Denote by $\{e_i\}_{i=1}^n$ the unit vectors in X and Y and by $\|\ \|_2$ the ℓ_2 norm of a vector. If

$$\max_{1\le i\le n} \|T^* e_i\|_2 \cdot \max_{1\le i\le n} \|T^{-1} e_i\|_2 = O(\log^\alpha n)$$

for some constant α, then

(35) $$\log_n \|T\|\ \|T^{-1}\| \ge \min_{0\le t\le 1} \operatorname{Osc}_{\sigma_t}\{f_Y(v) - f_X(u)\} + O(\log\log n/\log n).$$

In particular, if the distance between X and Y is attained (up to a logarithmic factor) by an orthogonal operator T, then problem 3 has a positive answer for such X and Y.

Proof. Proposition 3 follows directly from the proof of Corollary 3.6 of [L.S.]. We shall just point out how to derive (35) from that proof using the notation of [L.S.]. Let $\{a_{i,j}\}_{i,j=1}^n$ resp. $\{b_{i,j}\}_{i,j=1}^n$ be the matrix representations of T respectively T^{-1}. By the proof in [L.S.] there exist integers p,q and m so that if

$$\Delta = \{(i,j)\,;\, 2^p \le |a_{i,j}| \le 2^{p+1},\, 2^q \le |b_{i,j}| \le 2^{q+1}\}$$

and if $d = 2mn/|\Delta|$, then

$$|\{i\,;\, |\Delta_i| \ge m/d\}| \ge n/20\log^3 n,$$

$$|\{j\,;\, |\Delta^j| \ge n/d\}| \ge m/4,$$

$$2^{p+q} \ge d/(m\, 2\cdot 10^3 \log^5 n).$$

We clearly have

$$\max_i \|T^* e_i\|_2 \ge \sqrt{\frac{n}{d}}\, 2^p, \qquad \max_i \|T^{-1} e_i\| \ge \sqrt{\frac{n}{d}}\, 2^q.$$

Hence

$$\frac{n}{m}\ \frac{1}{2\cdot 10^3 \log^5 n} \le \frac{n}{d} 2^{p+q} \le C\ \log^\alpha n.$$

Consequently $s = \log_n m = 1 - O(\log\log n/\log n)$. As shown in [L.S.] we have for this s and a suitable z that

$$\log_n \|T\|\ \|T^{-1}\| \ge M(f_X, f_Y, z, s, 1) + M(-f_X, -f_Y, z, s, 1) + O(\log\log n/\log n).$$

Since $s = 1-O(\log\log n/\log n)$, replacing s by 1 produces another error term of size $O(\log\log n/\log n)$ and this yields (35).

□

We conclude this paper with an example of a pair of functions f and g satisfying (5) for which $\psi(f,g) > \theta(f,g) \vee \theta(g,f)$. Since by adding a constant to f or g we do not change $\psi(f,g)$ or $\theta(f,g)$, we can ignore the requirement $f(0) = g(0) = 0$ of (5).

We let $\varepsilon > 0$ and let f be defined by

$$f(0) = 5\varepsilon,\ f(0.2) = 4\varepsilon,\ f(0.3) = -5\varepsilon,\ f(0.72) = 4\varepsilon,\ f(1) = -2\varepsilon$$

and at the other values of t, the function $f(t)$ is determined so that it is as close to 0 as possible in view of the requirement that $|f(t)-f(s)| \le \frac{1}{2}|t-s|$ for all t,s. This means e.g. that f is linear with slope ½ from 0 to 10ε, f is equal to 0 from 10ε to $0.2-8\varepsilon$ and so on. Next we define g (taking $\delta \gg \varepsilon$)

$$g(t) = \begin{cases} -2\varepsilon & \text{for } 0 \le t \le 0.4-2\delta \\ 2\varepsilon & \text{for } 0.4-\delta \le t \le 0.4+\delta \\ \varepsilon & \text{for } 0.6-\delta \le t \le 0.6+\delta \\ -4\varepsilon & \text{for } 0.7-\delta \le t \le 0.7+\delta \\ 2\varepsilon & \text{for } 0.75-\delta \le t \le 0.75+\delta \\ \varepsilon & \text{for } t = 1 \end{cases}.$$

For other values of t the function g is again determined by the rule of being as close as possible to 0. A somewhat tedious but elementary computation shows that for small enough ε

$$\psi(f,g) = \min_{0 \le t \le 1} \operatorname*{Osc}_{\sigma_t}\{g(v)-f(u)\} \ge 10\varepsilon$$

while

$$\theta(f,g) \le M(f,g,0.25,1-2\varepsilon,1) + M(-f,-g,0.25,1-2\varepsilon,1) \le 9\varepsilon$$

$$\theta(g,f) \le M(f,g,0.6,1,1-2\varepsilon) + M(-f,-g,0.6,1,1-2\varepsilon) \le 9\varepsilon.$$

References

[B.M.] J. Bourgain and V. D. Milman, Distance between normed spaces, their subspaces and quotient spaces, Integral equations and operator theory 9(1986), 32-46.

[C.] S. Chevet, Séries de variables aléatoires Gaussiens à valeurs dans $E \otimes F$, Séminaire Maurey-Schwartz 1977-78, exp XIX.

[G.1] E. D. Gluskin, On the estimate of the distance between finite dimensional symmetric spaces, Iss. Lin. Oper. i Teorii Funk. 92 (1979), 268-273.

[G.2] E. D. Gluskin, On distances between some symmetric spaces, J. Soviet Math. 22 (1983), 1841-1846.

[G.G.] D. J. H. Garling and Y. Gordon, Relations between some constants associated with finite-dimensional Banach spaces, Israel J. Math. 9(1971), 356-361.

[J.M.S.T.] W. B. Johnson, B. Maurey, G. Schechtman and L. Tzafriri, Symmetric structures in Banach spaces, Memoirs A.M.S. 217, 1979.

[K] H. König, Spaces with large projection constants, Israel J. Math. 50 (1985), 181-188.

[L.S.] J. Lindenstrauss and A. Szankowski, The weak distance between Banach spaces with a symmetric basis, J. für die reine und angewandte Mathematik (Crelle J.) 373(1987), 108-147.

[R] S. Rolewicz, Metric linear spaces, Monografie Matematyczne, Warsaw 1972.

[S.1] C. Schütt, On the uniqueness of symmetric bases in finite dimensional Banach spaces, Israel J. Math. 40 (1981), 97-117.

[S.2] C. Schütt, Some geometric properties of finite dimensional symmetric Banach spaces, Séminaire d'Analyse Fonctionnelle Ecole Polytechnique, 1980/81 Exp 1.

[T.J.1] N. Tomczak Jaegermann, On the Banach Mazur distance between symmetric spaces, Bull. Acad. Polon. Sci. 27 (1979), 273-276.

[T.J.2] N. Tomczak Jaegermann, The weak distance between finite dimensional Banach spaces, Math. Nachr. 119 (1984), 291-307.

COMPLEMENTS OF SUBSPACES OF ℓ_p^n ; $p \geq 1$ WHICH ARE UNIQUELY DETERMINED

J. BOURGAIN and L. TZAFRIRI
IHES, France and University of Illinois   Hebrew University, Jerusalem

It is quite obvious that a complemented subspace X of a Banach space Z has a unique complement, up to isomorphism, since $Z = X \oplus Y$ implies that Y is isomorphic to the quotient space Z/X. The situation changes radically when we consider different complemented embeddings of X into Z. More precisely, $Z = X_1 \oplus Y_1 = X_2 \oplus Y_2$ with both X_1 and X_2 isomorphic to X does not necessarily imply that Y_1 is isomorphic to Y_2, as simple examples show. However, in the special case when $Z = \ell_p$, for some $p \geq 1$, we can conclude that Y_1 is isomorphic to Y_2, provided that they both have the same dimension. This fact is entirely trivial.

The purpose of the paper is to consider the local variant of this situation which is considerably more difficult and more interesting. The question of uniqueness of complements in $\ell_p^n : p \geq 1$, is also related to the still open problem whether the ℓ_p^n- spaces; $p \geq 1$, $n = 1,2,...$ are primary i.e. whether there exists a constant $M = M(p,K)$ so that, whenever $\ell_p^n = X \oplus Y$ and the norm of the projection P_X onto X which vanishes on Y is $\leq K$, then at least one of the factors, say X, satisfies $d(X, \ell_p^{\dim X}) \leq M$.

We present below a general principle which applies for certain spaces X and ensures the uniqueness, up to isomorphism, of the complement of any complementable embedding of X in ℓ_p^n. (Note that in the present context "isomorphic" means "well isomorphic" and "complemented" means "well complemented"; the precise meaning is explained in the sequel.) This principle is then used in two extremal cases.

The first and most interesting case occurs when $p > 1$ and X is a Hilbert space. We conclude that the situation $\ell_p^n = X \oplus Y$ with X isomorphic to Hilbert space determines uniquely the other factor Y, up to isomorphism. Actually, we also show that if dim X is smaller than the maximal dimension of a complemented Hilbert subspace of ℓ_p^n by a power of n then Y is even isomorphic to $\ell_p^{\dim Y}$.

Let us mention that factors of ℓ_p^n with small Hilbertian codimension arise naturally in analysis as e.g. complements of random Hilbert subspaces, L_Λ^p-spaces in harmonic analysis, etc. In these concrete settings, more abstract methods seems to be needed however in order to identify them.

The other case when we are able to prove the uniqueness of Y, up to isomorphism, takes place whenever X has a relatively small dimension; more exactly, when dim $X \leq c\sqrt{n}$, for some constant $c > 0$.

A BASIC RESULT

The basic result present below asserts, roughly speaking, that in the direct sum $\ell_p^n = X \oplus Y$, Y is uniquely determined provided there exists another decomposition $\ell_p^n = \ell_p^{\theta n} \oplus \ell_p^{(1-\theta)n}$, for some $0 < \theta < 1$, such that X embeds complementably in $\ell_p^{\theta n}$ and $\ell_n^{\theta n}$ is contained in Y. Before stating the theorem in precise terms, let us point out that the Banach-Mazur distance between two Banach spaces X and Y will be denoted by $d(X,Y)$. If $d(X,Y) = d$ we shall also put $X \overset{d}{\approx} Y$.

Theorem 1. Suppose that, for some $n,p \geq 1$ and $K \geq 1$, we have

$$\ell_p^n = X_1 \oplus Y_1 = X_2 \oplus Y_2$$

with $d(X_1,X_2) \leq K$. Assume also that there exists an integer m such that, for $i = 1,2$,

(i) $\ell_p^n = G_i \oplus H_i$ with $d(G_i, \ell_p^m) \leq K$, $d(H_i, \ell_p^{n-m}) \leq K$ and $G_i \subset Y_i$
(ii) $\ell_p^m = X' \oplus E$ with $d(X', X_i) \leq K$
(iii) all the projections determined by the above decompositions in direct sums have norms bounded by K.

Then $d(Y_1,Y_2) \leq 2^{14}K^{18}$.

Proof. Let $P_{X_i}, P_{Y_i}, P_{G_i}$ and P_{H_i} be the corresponding projections determined by the decompositions into direct sums $\ell_p^n = X_i \oplus Y_i$ and $\ell_p^n = G_i \oplus H_i$; $i = 1,2$. By (iii) and (i), we have for $x_i \in X_i$; $i = 1,2$,

$$K\,||x_i|| \geq ||P_{H_i}(x_i)|| = ||x_i - P_{G_i}(x_i)|| \geq ||x_i||/||P_{X_i}|| \geq ||x_i||/K$$

from which it follows that the subspaces $\tilde{X}_i = P_{H_i}(X_i)$ satisfy $d(X_i, \tilde{X}_i) \leq K^2$. We claim now that

$$\ell_p^n = \tilde{X}_i \oplus Y_i, \quad i = 1,2,$$

with the projections associated to these decompositions being bounded in norm by $2K^2$. Indeed, if $z \in \ell_p^n$ then $z = x_i + y_i$ with $x_i \in X_i$ and $y_i \in Y_i$, for $i = 1,2$. Thus, $z = P_{G_i}(x_i) + P_{H_i}(x_i) + y_i$ and, by (i), $P_{G_i}(x_i) + y_i \in Y_i$. Moreover,

$$||P_{H_i}(x_i)|| \leq K\,||x_i|| \leq K^2 ||z||.$$

and

$$||P_{G_i}(x_i)+y_i|| \leq ||P_{G_i}(x_i)|| + ||y_i|| \leq K\,||x_i|| + ||y_i|| \leq 2K^2 ||z||,$$

which proves our assertion.

The relative position of the decompositions $\ell_p^n = \tilde{X}_i \oplus Y_i$ and $\ell_p^n = G_i \oplus H_i$; $i = 1,2$, is explained by the following diagram

$$\begin{array}{c} G_i \\ \cap \\ \ell_p^n = \tilde{X}_i \oplus Y_i; \quad i = 1,2. \\ \cap \\ H_i \end{array}$$

Since $Y_i = G_i \oplus (Y_i \cap H_i)$ with the corresponding projections having norms bounded by K; $i = 1,2$, we can conclude that,

$$Y_i \overset{4K}{\approx} (G_i \oplus (Y_i \cap H_i))_p \overset{K}{\approx} (\ell_p^m \oplus (Y_i \cap H_i))_p \overset{1}{\approx}$$
$$((X' \oplus E) \oplus (Y_i \cap H_i))_p \overset{4K}{\approx} (X' \oplus E \oplus (Y_i \cap H_i))_p \overset{K}{\approx}$$
$$(X_i \oplus E \oplus (Y_i \cap H_i))_p \overset{K^2}{\approx} (\tilde{X}_i \oplus E \oplus (Y_i \cap H_i))_p \overset{8K^2}{\approx}$$
$$(E \oplus (\tilde{X}_i \oplus (Y_i \cap H_i)))_p \overset{1}{\approx} (E \oplus H_i)_p \overset{K}{\approx} (E \oplus \ell_p^{n-m})_p,$$

where the notation $(X \oplus Y)_p$ stands for the direct sum of X and Y in the sense of ℓ_p^2. It follows that

$$Y_i \overset{2^7K^9}{\approx} (E \oplus \ell_p^{n-m})_p; \quad i = 1,2,$$

from which we deduce the desired result. □

COMPLEMENTS OF HILBERTIAN SUBSPACES OF ℓ_p^n; $p > 1$.

The most interesting case in which Theorem 1 can be successfully applied is that of complemented Hilbertian subspaces of ℓ_p^n; $p > 1$.

Theorem 2. For every $p > 1$ and $K \geq 1$, there exists a constant $D = D(p,K)$ so that, whenever

$$\ell_p^n = X_1 \oplus Y_1 = X_2 \oplus Y_2,$$

for some n, and

(i) $\dim X_i = h$ and $d(X_i, \ell_2^h) \leq K$, for $i = 1,2$.

(ii) $\|P_{X_i}\|$, $\|P_{Y_i}\| \leq K$, for $i = 1,2$,

then $d(Y_1, Y_2) \leq D$.

Proof. Since the assertion is self-dual there is no loss of generality in assuming that $p > 2$. Fix $n, p > 2$ and $K \geq 1$, and consider a direct sum of the form $\ell_p^n = X \oplus Y$ with $\dim X = h$, $d(X, \ell_2^h) \leq K$ and such that $\|P_X\|, \|P_Y\| \leq K$. Let $\{e_i\}_{i=1}^n$ denote the unit vector basis of ℓ_p^n. Let $\alpha > 0$ and k be an integer which will be determined later. Denote by m the largest integer so that $km \leq n$ and suppose that

$$\|P_{X|[e_i]_{i=(j-1)m+1}^{jm}}\| \geq \alpha\,;\ j = 1,2,\ldots,k.$$

Then there exist vectors $u_j \in [e_i]_{i=(j-1)m+1}^{jm}$ such that $\|u_j\| = 1$ and $\|P_X(u_j)\| \geq \alpha$, for all $j = 1,2,\ldots,k$. Let $S : X \to \ell_2^h$ be an isomorphism so that $\|S\| = 1$ and $\|S^{-1}\| \leq K$. By Grothendieck's inequality in the formulation given in [3] (see also [4] II.1.f.14), it follows that

$$\|(\sum_{j=1}^{k} |SP_X(u_j)|^2)^{1/2}\| \leq K_G K \|(\sum_{j=1}^{k} |u_j|^2)^{1/2}\|.$$

Here K_G denotes the constant of Grothendieck and the absolute value $|u|$ of a vector $u = \sum_{i=1}^{n} a_i e_i$ is taken, as usual, to be $|u| = \sum_{i=1}^{n} |a_i| e_i$. Similarly, $|u|^2$ stands for $\sum_{i=1}^{n} |a_i|^2 e_i$.

Since $\{u_j\}_{j=1}^{k}$ are mutually disjoint vectors in ℓ_p^n we conclude that

$$\alpha k^{1/2} K^{-1} \leq K_G K\, k^{1/p}$$

i.e.

$$\alpha \leq K_G K / k^{1/2-1/p}.$$

Suppose now that k was chosen to be the smallest integer which contradicts the above inequality i.e.

$$k > (K_G K^2/\alpha)^{2p/(p-2)}.$$

Then there exists at least one value $1 \leq j_0 \leq k$ of j such that

$$\|P_{X|[e_i]_{i=(j_0-1)m+1}^{j_0 m}}\| < \alpha.$$

Put $\sigma = \{(j_0-1)m + 1,\ (j_0-1)m + 2, \ldots, j_0 m\}$. It follows from the above inequality that $\{e_i\}_{i\in\sigma}$ is $K(1-\alpha)^{-1}$-equivalent to $\{P_Y(e_i)\}_{i\in\sigma}$ and, moreover, that the projection R from ℓ_p^n onto $\{P_Y(e_i)\}_{i\in\sigma}$ which vanishes on $e_i : i \notin \sigma$ satisfies

$$\|R\| \leq 2K(1-4\alpha K)^{-1},$$

provided of course that $0 < \alpha < 1/4K$. We can now determine α and assume that it was chosen small enough as to ensure that $(1-\alpha)^{-1}$, $(1-4\alpha K)^{-1} \leq 2$. Then, $\ell_p^n = G \oplus H$ with $G = [P_Y(e_i)]_{i\in\sigma} \subset Y$, $H = [e_i]_{i\notin\sigma}$, $d(G,\ell_p^m) \leq 2K$, $d(H,\ell_p^{n-m}) = 1$ and the projections associated to this direct sum have norms $\leq 5K$.

In order to be able to use Theorem 1 and complete the proof of Theorem 2, it remains to show that ℓ_p^m contains a well-complemented copy of X i.e. of ℓ_2^h. To this purpose, notice that $km + k > n$ i.e.

$$m > \frac{n}{k} - 1 > \frac{n}{[(K_G K^2/\alpha)^{2p(p-2)}] + 1} - 1$$

or, in other words, $m \geq \beta n$, for some $\beta = \beta(p,K)$ which satisfies $0 < \beta < 1$.

However, it is well-known (cf. [1], [2]) that, for every $p > 2$ and $K \geq 1$, there exist constants $c_1(p,K)$ and $c_2(p,K)$ such that, for all n, we have that

1° ℓ_p^n contains a subspace F_0 of dimension $s_0 \geq c_1(p,K)n^{2/p}$ so that $d(F_0,\ell_2^{s_0}) \leq K$

2° every subspace F of ℓ_p^n of dimension s for which $d(F,\ell_2^s) \leq K$ must satisfy $s \leq c_2(p,K)n^{2/p}$.

The argument from [2] shows actually that

$$c_1(p,K)\ell^{1-2/p} \leq c_2(p,K\ell^{1/2-1/p}),\ \ell = 1,2,\cdots .$$

This fact can be also deduced from 1° and 2° in a formal manner, as follows. Fix ℓ, choose a large k and put $n = k\ell$. By 1°, ℓ_p^k contains a subspace F_k of dimension s_k so that $s_k \geq c_1(p,K)k^{2/p}$ and $d(F_k,\ell_2^{s_k}) \leq K$. Hence, ℓ_p^n contains the subspace $F = \underbrace{(F_k \oplus F_k \oplus \cdots \oplus F_k)_p}_{\ell \text{ times}}$ of dimension $s = \ell s_k \geq c_1(p,K)k^{2/p}\cdot\ell$ which clearly satisfies $d(F,\ell_2^s) \leq K\ell^{1/2-1/p}$. It follows from 2° that

$$c_1(p,K)k^{2/p}\,\ell \leq c_2(p,K\ell^{1/2-1/p})n^{2/p},$$

which clearly implies the inequalities above.

In order to prove that ℓ_p^m contains an isomorphic copy of ℓ_2^h, notice that, by (i), $h \leq c_2(p,K)n^{2/p}$. Choose now ℓ so that

$$c_2(p,K)/c_1(p,K) \leq \ell^{1-2/p}\beta^{2/p} \leq 2c_2(p,K)/c_1(p,K).$$

Then

$$h \leq c_1(p,K)\ell^{1-2/p}\beta^{2/p}n^{2/p} \leq c_2(p,K\ell^{1/2-1/p})m^{2/p}$$

which, by the above argument, implies that ℓ_p^m contains indeed a subspace X_0 so that $d(X_0,\ell_2^h) \leq K\ell^{1/2-1/p} \leq K\beta^{-1/p}(2c_2(p,K)/c_1(p,K))^{1/2} = K_0$. The proof is then completed if we recall that, by [5] Corollary. 79, X_0 is $K_0\sqrt{p}$-complemented in ℓ_p^m. □

It is very likely that, whenever ℓ_p^n is isomorphic to $\ell_2^h \oplus Y$, then Y is well-isomorphic to ℓ_p^{n-k}. Some progress in this direction was achieved in [1] where it was shown that, for every $p > 1$, there exists a constant $M = M(p)$ such that ℓ_p^n is M-isomorphic to $\ell_2^k \oplus \ell_p^{n-k}$, provided that $k \leq n^{2/(p+2)}$. We shall prove in the sequel that this result remains valid for $k \leq n^{2/p-\varepsilon}$, where ε is an arbitrary positive number. We need first the following lemma which is, in fact, a generalization of the argument in [1].

Lemma 3. For every $p > 2$ and $L \geq 1$, there exists a constant $M = M(p,L)$ such that, for any n and $k < n^{2/p}$, whenever ℓ_p^n is L-isomorphic to $(\ell_2^k \oplus \ell_p^{n-k})_p$ and m is an integer for which $km \leq n^{2/p}$, then ℓ_p^{nm} is M-isomorphic to $(\ell_2^{km} \oplus \ell_p^{(n-k)m})_p$.

Proof. Suppose that ℓ_p^n is L-isomorphic to $(\ell_2^k \oplus \ell_p^{n-k})_p$, for some $p > 2$, $L \geq 1$, n and $k < n^{2/p}$. Let m be an integer satisfying $km \leq n^{2/p}$. Then ℓ_p^{nm} is L-isomorphic to the direct sum in the sense of ℓ_p^m of m copies of $(\ell_2^k \oplus \ell_p^{n-k})_p$. Since $k(m-1) \leq (n-k)^{2/p}$ it follows from 1° (in the proof of Theorem 2 above) that there exist a $K = K(p)$ and a sequence of spaces

$\{F_i\}_{i=1}^{m}$ such that, for each $1 \le i \le m$, ℓ_p^{n-k} is K-isomorphic to $(\ell_2^{k(m-i)} \oplus F_i)_p$. Thus, ℓ_p^{nm} is KL- isomorphic to the direct sum

$$(\ell_2^k \oplus \ell_2^{k(m-1)} \oplus F_1 \oplus \ell_2^k \oplus \ell_2^{k(m-2)} \oplus F_2 \oplus \cdots \oplus \ell_2^k \oplus \ell_2^0 \oplus F_m)_p.$$

However, each of the direct sums

$$(\ell_2^k \oplus \ell_2^{k(m-2)} \oplus F_1)_p, (\ell_2^k \oplus \ell_2^{k(m-3)} \oplus F_2)_p, \ldots, (\ell_2^k \oplus \ell_2^0 \oplus F_{m-1})_p,$$

as well as F_m, is $2^{1/2-1/p} K$-isomorphic to ℓ_p^{n-k}. Hence, ℓ_p^{nm} is $2^{1/2-1/p}$ KL-isomorphic to $(\ell_2^{km} \oplus \ell_p^{(n-k)m})_p$. □

Theorem 4. For every $p > 1$, $K \ge 1$ and $\varepsilon > 0$, there exists a constant $S = S(p,K,\varepsilon)$ so that, whenever ℓ_p^n is K-isomorphic to the direct sum $(\ell_2^k \oplus Y)_p$, for some space Y and integers n and $k \le n^{2/p-\varepsilon}$, then $d(Y, \ell_p^{n-k}) \le S$.

Proof. By duality, we can assume without loss of generality that $p > 2$. Fix integers n and $k \le n^{2/p-\varepsilon}$, for some $\varepsilon > 0$. Let $h = n/k$ and suppose that h is an integer (we shall omit the case when h is not an integer which can be completed easily by the reader). Since

$$h \ge n^{1-2/p+\varepsilon}$$

we get that

$$h^{\sum_{i=0}^{\infty}(2/p)^i} \ge n^{1+\varepsilon/(1-2/p)}.$$

Hence, we can find an integer j so that

$$h^{\sum_{i=0}^{j}(2/p)^i} = n$$

(again, we omit the discussion concerning the fact that, in general, we do not obtain exactly n).

We begin now an inductive procedure involving the use of Lemma 3. We first notice the trivial fact that ℓ_p^h is isometric to $(\ell_2^1 \oplus \ell_p^{h-1})_p$. By applying Lemma 3 with $m = h^{2/p}$, we conclude that $\ell_p^{h^{1+2/p}}$ is $M_1 = M(p,1)$-isomorphic to the direct sum $(\ell_2^{h^{2/p}} \oplus \ell_p^{s_1})_p$, for a suitable integer s_1 (recall that $M(p,L)$ is the constant appearing in the statement of Lemma 3). By repeating the argument with $m = h^{(2/p)^2}$, we get that $\ell_p^{h^{1+2/p+(2/p)^2}}$ is $M_2 = M(p, M_1)$-isomorphic to $(\ell_2^{h^{2/p+(2/p)^2}} \oplus \ell_p^{s_2})_p$, for a suitable s_2. Continuing j times, we obtain the existence of a constant $M_j = M_j(p)$ such that ℓ_p^n is M_j-isomorphic to the direct sum

$$(\ell_2^{h^{\sum_{i=1}^{j}(2/p)^i}} \oplus \ell_p^{s_j})_p,$$

for a suitable s_j, i.e. that ℓ_p^n is M_j- isomorphic to $(\ell_2^k \oplus \ell_p^{n-k})_p$. The final conclusion follows immediately from Theorem 2. □

COMPLEMENTS OF SUBSPACES OF $\ell_p^n; p \ge 1$, OF SMALL DIMENSION

The purpose of this section is to prove that a subspace X of ℓ_p^n has a uniquely determined complement when its dimension does not exceed $cn^{1/2}$, for some constant c. The restriction imposed on the dimension of X can be relaxed on the extent the type 2 constant $T_2(X)$ of X satisfies some non-trivial estimates.

Theorem 5. For every $K \geq 1$, there exist constant $c = c(K) > 0$ and $C = C(K) < \infty$ such that, whenever

$$\ell_p^n = X_1 \oplus Y_1 = X_2 \oplus Y_2,$$

for some integer n and some $1 \leq p \leq \infty$, and

(i) $\dim X_1 = \dim X_2 = h$ and $d(X_1, X_2) \leq K$,
(ii) $\|P_{X_i}\|, \|P_{Y_i}\| \leq K$; $i = 1,2$,
(iii) $h \leq cn^{1/2}$

or

(iii)' $hT_2(\hat{X}_i)^{|1/p-1/2|^{-1}} \leq cn$, $i = 1,2$,

where $\hat{X}_i = X_i$ if $1 \leq p \leq 2$ or $\hat{X}_i = X_i^*$ when $p > 2$, then

$$d(Y_1, Y_2) \leq C.$$

The proof of Theorem 5 is, as easily verified, an immediate consequence of Theorem 1 and the following two propositions, each of which is of interest in itself.

Proposition 6. For every $K \geq 1$, there exist constants $d = d(K) > 0$. $D = D(K) < \infty$ and $0 < \beta = \beta(K) < 1$ so that, whenever $\ell_p^n = X \oplus Y$, for some n and $1 \leq p \leq 2$, and

(i) $\|P_X\|$, $\|P_Y\| \leq K$,
(ii) $\dim X \leq dn^{1/2}$

or

(ii)' $(\dim X)T_2(X)^{(1/p-1/2)^{-1}} \leq dn$

then there exist an integer $s \geq \beta n$ and subspace G and H of ℓ_p^n so that $\ell_p^n = G \oplus H$, $d(G, \ell_p^s) \leq D$, $d(H, \ell_p^{n-s}) \leq D$, $G \subset Y$ and $\|P_G\|, \|P_H\| \leq D$ (P_G and P_H are the projections associated to the direct sum $\ell_p^n = G \oplus H$).

Proposition 7. For every $K \geq 1$ and $0 < \beta < 1$, there exist constants $b = b(K,p) > 0$ and $B = B(K,p)$ such that, whenever $\ell_p^n = X \oplus Y$, for some n and $1 \leq p \leq 2$, and

(i) $\|P_X\|, \|P_Y\| \leq K$,
(ii) $\dim X \leq bn^{1/2}$

or

(ii') $(\dim X)T_2(X)^{(1/p-1/2)^{-1}} \leq bn$,

then, for each integer $r \geq \beta n$, we have $\ell_p^r = X' \oplus E$ with $d(X', X) \leq B$ and $\|P_{X'}\|, \|P_E\| \leq B$.

The proofs of Propositions 6 and 7 rely on the following lemma.

Lemma 8. For each $0 < \tau < 1$, there exist constants $A = A(\tau)$ and $\varepsilon = \varepsilon(\tau)$ so that the following holds. Let $1 \leq p \leq 2$, let h, n and m be integers for which $m \geq A(\tau)h$ and assume that X is a subspace of ℓ_p^n of dimension h and I a subset of the integers $\{1, 2, \ldots, n\}$ such that, for every $x \in X$ of norm one and $J \subset I$ of cardinality $|J| \leq m$, we have $\|R_J(x)\| \leq \varepsilon$, where R_J denotes the norm one projection on ℓ_p^n, defined by $R_J(\sum_{j=1}^{n} a_j e_j) = \sum_{j \in J} a_j e_j$. Let $\{\varphi_j\}_{j \in I}$ be a sequence of independent random variables of mean zero over a probability space (Ω, Σ, μ) taking only two values $1-\delta$ and $-\delta$, for some $0 < \delta < 1$. Then

$$S = \int_\Omega \sup\{|\sum_{j \in I} |a_j|^p \varphi_j(\omega)| \; ; x = \sum_{j=1}^{n} a_j e_j \in X, \; \|x\| \leq 1\} d\mu < \tau.$$

Proof. Take $\varepsilon = \tau/5$ and let $\mathcal{E}$ be an ε-net in the unit ball of X whose cardinality ℓ satisfies

$$\ell \leq (2/\varepsilon)^h.$$

For each $x = \sum_{j=1}^{n} a_j e_j \in \mathcal{E}$ put

$$J_x = \{j \in I : |a_j|^p \leq 1/m\}.$$

Then $1 \geq \|x\|^p \geq |I \sim J_x|/m$ i.e.

$$|I - J_x| \leq m.$$

Hence, by our assumption, $\|R_{I \sim J_x}(x)\| \leq \varepsilon$. It follows that, for each $\omega \in \Omega$ and $x = \sum_{j=1}^{n} a_j e_j \in \mathcal{E}$ we have

$$|\sum_{j \in I} |a_j|^p \varphi_j(\omega)| \leq \varepsilon^p + |\sum_{j \in J_x} |a_j|^p \varphi_j(\omega)|.$$

Therefore,

$$S \leq \varepsilon^p + \int_\Omega \max\{|\sum_{j \in I} |a_j|^p \varphi_j(\omega)|; x = \sum_{j=1}^{n} a_j e_j \in \mathcal{E}\} d\mu \leq$$

$$\leq 2\varepsilon^p + \int_\Omega \max\{|\sum_{j \in J_x} |a_j|^p \varphi_j(\omega)| \; ; x = \sum_{j=1}^{n} a_j e_j \in \mathcal{E}\} d\mu.$$

However, the sequence $\{\varphi_j\}_{j \in I}$ satisfies Khintchine's inequality in the sense that, for every integer k and every sequence $\{b_j\}_{j \in I}$ of scalars, we have

$$\int_\Omega |\sum_{j \in I} b_j \varphi_j(\omega)|^{2k} d\mu \leq 2\delta(1-\delta) k^k (\sum_{j \in I} |b_j|^2)^k$$

from which we deduce the existence of an absolute constant $\gamma > 0$ (independent of δ) such that

$$\int_{\Omega} e^{\gamma|\sum_{j\in I} b_j\varphi_j(\omega)|^2} d\mu \leq 1,$$

whenever $\sum_{j\in I} |b_j|^2 \leq 1$. Therefore, for $\{b_j\}_{j\in I}$ as above and any $\lambda > 0$, we get

$$\mu\{\omega\in\Omega\,;\, |\sum_{j\in I} b_j\varphi_j(\omega)| > \lambda\} < e^{-\gamma\lambda^2}.$$

Now, note that, for $x = \sum_{j=1}^{n} a_j e_j \in \mathcal{E}$, we have

$$\sum_{j\in J_x} |a_j|^{2p} \leq \sum_{j\in J_x} |a_j|^p/m \leq 1/m.$$

Hence,

$$\mu\{\omega\in\Omega;\, \max[m^{1/2}|\sum_{j\in J_x} |a_j|^p\varphi_j(\omega)|\,;\, x = \sum_{j=1}^{n} a_j e_j \in \mathcal{E}] > \lambda\} \leq$$

$$\leq \Sigma[\mu\{\omega\in\Omega;\, m^{1/2}|\sum_{j\in J_x} |a_j|^p\varphi_j(\omega)| > \lambda\}\,;\, x = \sum_{j=1}^{n} a_j e_j \in \mathcal{E}] \leq \ell e^{-\gamma\lambda^2},$$

which implies that

$$S \leq 2\varepsilon^p + \varepsilon + \int_{\varepsilon}^{\infty} \mu\{\omega\in\Omega;\, \max[|\sum_{j\in J_x} |a_j|^p\varphi_j(\omega)|;\, x = \sum_{j=1}^{n} a_j e_j \in \mathcal{E}] > \lambda\} d\lambda \leq$$

$$\leq 3\varepsilon + \ell m^{-1/2} \int_{\varepsilon m^{1/2}}^{\infty} e^{-\gamma\lambda^2} d\lambda \leq 3\varepsilon + \ell m^{-1/2} e^{-\gamma m\varepsilon^2} L,$$

where $L = \int_0^{\infty} e^{-\gamma\lambda^2} d\lambda$. Let now

$$A(\tau) = \max\{\log(2/\varepsilon)/\gamma\varepsilon^2, L^2/\varepsilon^2\}$$

and suppose that the condition $m \geq A(\tau)h$ holds. Then

$$S \leq 3\varepsilon + m^{-1/2} L\, e^{\log \ell - \gamma m\varepsilon^2} \leq 3\varepsilon + m^{-1/2} L\, e^{h\log(2/\varepsilon) - \gamma m\varepsilon^2}$$

$$\leq 3\varepsilon + m^{-1/2} L \leq 4\varepsilon < \tau,$$

which completes the proof. □

We can present now the proof of Proposition 6. Fix $K \geq 1$ and suppose that $\ell_p^n = X \oplus Y$, for some $1 \leq p \leq 2$ and integer n, so that $\|P_X\|, \|P_Y\| \leq K$. Take $\tau = 1/2(4K)^{2p}$ and let $A(\tau)$ and $\varepsilon(\tau)$ be given by Lemma 8. Then with the notation $h = \dim X$ take $m = [A(\tau)h] + 1$ and construct inductively a maximal set of mutually disjoint subsets $\{I_j\}_{j=1}^{k}$ of the integers $\{1,2,\ldots,n\}$ and norm one vectors $\{x_j\}_{j=1}^{k}$ in X such that $|I_j| \leq m$ and $\|R_{I_j}(x_j)\| > \varepsilon(\tau)$, for all $1 \leq j \leq k$. If $I = \{1,2,\ldots,n\} \sim \bigcup_{j=1}^{k} I_j$ and J is a subset of I with $|J| \leq m$ then, by the above maximality, $\|R_J(x)\| \leq \varepsilon(\tau)\|x\|$; $x \in X$, which shows that m and I satisfy the conditions required by Lemma 8.

Take $\delta = 1/(4K)^p$ and let $\{\xi_j\}_{j\in I}$ be a sequence of independent random variables of mean δ over a probability space (Ω,Σ,μ), each taking only the values 0 and 1. For $\omega \in \Omega$, define

$$I_\omega = \{j \in I \,;\, \xi_j(\omega) = 1\}$$

and note that

$$\int_\Omega |I_\omega| \, d\mu = \int_\Omega \sum_{j\in I} \xi_j(\omega) d\mu = \delta |I|.$$

On the other hand, by Lemma 8 applied with $\varphi_j(\omega) = \xi_j(\omega) - \delta$, $j \in I$,

$$\int_\Omega \sup\{|\sum_{j\in I} |a_j|^p (\xi_j(\omega) - \delta)| \,;\, x = \sum_{j=1}^n a_j e_j \in X; \|x\| \le 1\} d\mu < \tau.$$

A simple probability computation shows that there exists a $\omega_0 \in \Omega$ so that

$$|I_{\omega_0}| > \delta |I|/3 = |I|/3(4K)^p$$

and

$$S_0 = \sup\{|\sum_{j\in I} |a_j|^p (\xi_j(\omega_0) - \delta)| ; x = \sum_{j=1}^n a_j e_j \in X; \|x\| \le 1\} \le 1/(4K)^p.$$

Select now a vector $z = \sum_{j=1}^n c_j e_j$ in X such that $\|z\| \le \|P_X\|$ and

$$\|R_{I_\omega} P_X\| = \|R_{I_\omega}(z)\|.$$

Then

$$\|R_{I_\omega} P_X\|^p = \sum_{j\in I} |c_j|^p \xi_j(\omega_0) \le \delta \|z\|^p + |\sum_{j\in I} |c_j|^p (\xi_j(\omega_0) - \delta)| \le$$

$$\le \delta K^p + S_0 K^p \le 1/2^p,$$

i.e.

$$\|R_{I_\omega} P_X\| \le 1/2.$$

Put $G = P_Y R_{I_\omega} \ell_p^n$, $H = [e_j]_{j \notin I_\omega}$ and note that, for $u \in R_{I_\omega} \ell_p^n$ and v belonging to H,

$$\|P_Y(u) + v\| = \|u + v - P_X(u)\| \ge \|u - R_{I_\omega} P_X(u)\| \ge$$

$$\ge \|u\| - \|R_{I_\omega} P_X(u)\| \ge \|u\|/2.$$

This implies that

$$K\|u\| \ge \|P_Y(u)\| \ge \|u\|/2$$

i.e. $d(G, \ell_p^s) \le 2K$, where $s = |I_{\omega_0}|$. Furthermore, we also get that $\|P_G\| \le 2K$ and $\|P_H\| \le 1 + 2K$.

The proof would be now completed provided we prove that $s = |I_{\omega_0}|$ satisfies the required estimate from below under the assumption that dim X is not too large.

We first need to estimate k. Since

$$\varepsilon(\tau)k^{1/p} \le \| \sum_{j=1}^{k} R_{I_j}(x_j) \| \le \| \max_{1\le j\le k} |x_j| \| \le \| (\sum_{j=1}^{k} |x_j|^2)^{1/2} \| \le$$

$$\le \sqrt{2} \int_0^1 \| \sum_{j=1}^{k} r_j(t) x_j \| dt \le \sqrt{2} T_2(X) k^{1/2}$$

we obtain that

$$k \le (\sqrt{2}\, T_2(X)/\varepsilon(\tau))^{(1/p-1/2)^{-1}}.$$

However, by using [6], we easily check that

$$T_2(X) \le \sqrt{2}\, h^{1/p-1/2}$$

which yields that

$$k \le (2/\varepsilon(\tau))^{1/p-1/2} h.$$

It follows that

$$km \le \min(2A(\tau)(2/\varepsilon(T))^{1/p-1/2} h^2, 2A(\tau)(\sqrt{2}/\varepsilon(\tau))^{1/p-1/2} h\; T_2(X)^{(1/p-1/2)^{-1}})$$

and we can ensure that $km \le n/2$ for a suitable choice of d. This yields that

$$|I_{\omega_0}| \ge (n-km)/3(4K)^p \ge n/6(4K)^p,$$

thus completing the proof (since, as easily verified, the constants d, D and β can be chosen as not to depend on p). □

Before presenting the proof of Proposition 7, we need the following lemma.

Lemma 9. For every $K > 1$ and $0 < \alpha < 1$, there exists a constant $a = a(K,\alpha)$ such that, whenever $\ell_p^n = X \oplus Y$, for some n and $1 \le p \le 2$, and

(i) $\|P_X\|, \|P_Y\| \le K$,
(ii) $\dim X \le a n^{1/2}$

or

(ii') $(\dim X) \cdot T_2(X)^{(1/p-1/2)^{-1}} \le an$,

then one can find a subset I' of the integers $\{1,2,\ldots,n\}$ of cardinality $|I'| \le [\alpha n]$ for which

$$\|x\| \ge \|R_{I'}(x)\| \ge \alpha^{1/p} \|x\|/8;\ x \in X$$

(in other words, X is $8/\alpha^{1/p}$-isomorphic to a subspace of $\ell_p^{[\alpha n]}$).

Proof. We begin exactly as in the proof of Proposition 6 except that we take $\tau = \alpha^2/64.4^p$. Let $\{I_j\}_{j=1}^k, I, m, \{\xi_j\}_{j\in I}$ and Ω have the same meaning as above. Then, for $\delta = \alpha/2\cdot 2^p$, a similar random argument using Lemma 8 shows that there exists a $\omega_1 \in \Omega$ such that

$$|I_{\omega_1}| < 2\delta |I| \le \alpha n/2$$

and

$$S_1 = \sup\{ |\sum_{j\in I} |a_j|^p (\xi_j(\omega_1) - \delta)| \; ; x = \sum_{j=1}^{n} a_j e_j \in X, \; \|x\| \le 1\} \le \alpha/4\cdot 2^p .$$

Put

$$I' = (\bigcup_{j=1}^{k} I_j) \cup I_{\omega_1}$$

and note that, for any $\tilde{x} = \sum_{j=1}^{n} c_i e_i \in X$ with $\|\tilde{x}\| = 1$, we have

$$\|R_{I'}(\tilde{x})\|^p = \sum_{i\in \bigcup_{j=1}^{k} I_j} |c_i|^p + \sum_{i\in I} |c_i|^p \xi_i(\omega_1) = \sum_{i\in \bigcup_{j=1}^{k} I_j} |c_i|^p + \delta \sum_{i\in I} |c_i|^p +$$

$$+ \sum_{i\in I} |c_i|^p (\xi_i(\omega_1) - \delta) \ge \delta - S_1 \ge \alpha/4\cdot 2^p$$

i.e. $\|R_{I'}(\tilde{x})\| \ge \alpha^{1/p}/8$.It remains to evaluate the cardinality of I' which clearly satisfies

$$|I'| \le km + |I_{\omega_1}| \le km + \alpha n/2.$$

The proof can now be completed by ensuring that $km \le \alpha n/2$. This can be done by choosing a suitable a in view of the conditions (ii) or (ii') and the estimate given for km at the end of the proof of Proposition 6. □

Proof of Proposition 7. Fix $1 \le p \le 2$ and n, and suppose that we have the direct decomposition $\ell_p^n = X \oplus Y$ so that condition (i) of Proposition 7 holds. Fix $0 < \alpha < 1$ and apply Lemma 9. Then if (ii) or (ii') of Lemma 9 are satisfied for the corresponding $a = a(K,\alpha)$ we can find a subset I' of $\{1,2,\dots,n\}$ so that $|I'| \le [\alpha n]$ and

$$\|R_{I'}(x)\| \ge \alpha^{1/p} \|x\|/8; \; x \in X.$$

Take $\tau = \alpha^2/64\cdot 4^p K$ and let $\varepsilon = \varepsilon(\tau)$ and $A(\tau)$ be given by Lemma 8. Put $m = [A(\tau)\dim X] + 1$ and notice that, by the construction of I', we have

$$\|R_J(x)\| \le \varepsilon \|x\|,$$

for any $x \in X$ and any subset J of $I'' = \{1,2,\dots,n\} \sim I'$ of cardinality $\le m$.

Let $\{\psi_j\}_{j\in I''}$ be a sequence of independent random variables of mean α over a probability space (Ω,Σ,μ) which take only the values 0 and 1. For each $\omega \in \Omega$, put

$$I''_{\omega} = \{j \in I'' \; ; \psi_j(x) = 1\}$$

and $\bar{I}_{\omega} = I' \cup I''_{\omega}$.

Define now a diagonal operator D_{ω} on $R_{\bar{I}_\omega}\ell_p^n$, by setting $D_{\omega}(e_i) = 1$ if $i \in I'$ and $D_{\omega}(e_i) = \alpha^{-1}$ if $i \in I''_{\omega}$. Furthermore, for every $x = \sum_{j=1}^{n} a_j e_j \in X$ with $\|x\| = 1$, choose a functional

$x^* = \sum_{j\in I'} b_j e_j^* \in (\ell_p^n)^*$ so that $\|x^*\| = 1, x^*(x) \geq \alpha^{1/2}$.

Hence, if we denote by $(p_{ij})_{i,j=1}^n$ the matrix associated to the operator P_X (relative to the usual basis $\{e_i\}_{i=1}^n$ of ℓ_p^n) then

$$\|R_{\bar{I}_\omega} P_X D_\omega R_{\bar{I}_\omega}(x)\| \geq \|R_{I'} P_X D_\omega R_{\bar{I}_\omega}(x)\| =$$

$$= \|\sum_{\ell\in I'} (\sum_{j\in I'} a_j p_{j\ell} + \alpha^{-1} \sum_{j\in I_\omega^\cdot} a_j p_{j\ell}) e_\ell\| \geq$$

$$\geq x^*(\sum_{\ell\in I'} (\sum_{j\in I'} a_j p_{j\ell} + \alpha^{-1} \sum_{j\in I_\omega^\cdot} a_j p_{j\ell}) e_\ell)$$

$$= \sum_{\ell\in I'} b_\ell (\sum_{j\in I'} a_j p_{j\ell} + \alpha^{-1} \sum_{j\in I_\omega^\cdot} a_j p_{j\ell}) = x^*(P_X(x)) +$$

$$+ \alpha^{-1} \sum_{j\in I^\cdot} (\sum_{\ell\in I'} b_\ell p_{j\ell}) a_j (\psi_j(\omega) - \alpha) \geq$$

$$\geq \alpha^{1/2} - \alpha^{-1} |\sum_{j\in I^\cdot} (\sum_{\ell\in I'} b_\ell p_{j\ell}) a_j (\psi_j(\omega) - \alpha)|.$$

Consider now the average

$$S' = \int_\Omega \sup\{|\sum_{j\in I^\cdot} (\sum_{\ell\in I'} d_\ell p_{j\ell}) c_j (\psi_j(\omega) - \alpha)|;$$

$$x = \sum_{j=1}^n c_j e_j \in X, x^* = \sum_{\ell\in I'} d_\ell e_\ell^* \in (\ell_p^n)^*\ ;\ \|x^*\| \leq 1, x^*(x) \geq \alpha^{1/2}\} d\mu$$

and note that, by Holder's inequality, we obtain

$$\sum_{j\in I^\cdot} |(\sum_{\ell\in I'} d_\ell p_{j\ell}) c_j| \leq (\sum_{j\in I^\cdot} |c_j|^p)^{1/p} \cdot (\sum_{j\in I^\cdot} |\sum_{\ell\in I'} d_\ell p_{j\ell}|^2)^{1/2} \leq \|P_X^*(x^*)\| \leq K.$$

Hence, by Lemma 8 applied with ψ_j instead of ξ_j and $|(\sum_{\ell\in I'} d_\ell p_{j\ell}) c_j|/K$ instead of $|a_j|^p$, we get that

$$S' \leq K\tau.$$

On the other hand,

$$\int_\Omega |\bar{I}_\omega| d\mu = |\bar{I}| + \int_\Omega \sum_{j\in I^\cdot} \psi_j(\omega) d\mu < 2\alpha n.$$

Therefore, as is readily verified, there exists a $\rho \in \Omega$ such that

$$|\bar{I}_\rho| \leq 4\alpha n$$

and

$$\| R_{\bar{I}_\rho} P_X D_\rho R_{\bar{I}_\rho(x)} \| \geq \alpha^{1/2} \| x \| / 2, \quad x \in X.$$

We are ready now to complete the proof. Put

$$\bar{X} = D_\rho R_{\bar{I}_\rho} X$$

and note that $\bar{X}$ is a subspace of $[e_i]_{i \in \bar{I}_\rho}$ i.e. a subspace of $\ell_p^{|\bar{I}_\rho|}$. Moreover, for every $x \in X$,

$$\alpha^{-1} \| x \| \geq \| D_\rho R_{\bar{I}_\rho}(x) \| \geq K^{-1} \| R_{\bar{I}_\rho} P_X D_\rho R_{\bar{I}_\rho}(x) \| \geq (2K)^{-1} \alpha^{1/2} \| x \|$$

i.e. $d(X, \bar{X}) \leq 2K / \alpha^{3/2}$. Let $T : [e_i]_{i \in \bar{I}_\rho} \to \bar{X}$ be an operator defined by

$$T(z) = D_\rho R_{\bar{I}_\rho} P_X(z) : z \in [e_i]_{i \in \bar{I}_\rho}$$

and take a vector $z = D_\rho R_{\bar{I}_\rho}(x) \in \bar{X}$, for a suitable $x \in X$. Then

$$\| T(z) \| = \| D_\rho R_{\bar{I}_\rho} P_X D_\rho R_{\bar{I}_\rho}(x) \|$$

$$\geq \| R_{\bar{I}_\rho} P_X D_\rho R_{\bar{I}_\rho}(x) \| \geq \alpha^{1/2} \| x \| / 2 \geq \alpha^{3/2} \| z \| / 2$$

which shows that $T_{|\bar{X}}$ is an invertible operator of norm $\leq 2/\alpha^{3/2}$. It follows immediately that

$$Q = (T_{|\bar{X}})^{-1} T$$

is a projection of norm $\leq 2K/\alpha^{5/2}$ from $[e_i]_{i \in \bar{I}_\rho}$ onto $\bar{X}$, thus completing the proof since the cardinality of $\bar{I}_\rho$ satisfies $|\bar{I}_\rho| \leq 4\alpha n$ and α can be taken as small as possible. □

REFERENCES

1. G. Bennett, L. E. Dor, V. Goodman, W. B. Johnson and C. M. Newman, On uncomplemented subspaces of $L^p, 1 < p < 2$, Israel J. Math., 26 (1977), 178-187.
2. T. Figiel, J. Lindenstrauss and V. Milman, The dimension of almost spherical sections of convex bodies, Acta Math., 139 (1977), 53-94.
3. J. L. Krivine, Théorèmes des factorisation dans les espaces réticulés, Séminaire Maurey-Schwartz, 1973-74, Exposés 23-24, Ecole Polyt. Paris.
4. J. Lindenstrauss and L. Tzafriri, Classical Banach spaces II, Function spaces, Springer Verlag, Berlin 1979.
5. B. Maurey, Théorèmes de factorisation pour les opérateurs à valeurs dans les espaces L^p, Astérisque, Soc. Math. France 11 (1974)
6. N. Tomczak-Jaegermann, Computing 2-summing norm with few vectors, Ark. Mat. 17(1979), 173-177.

EMBEDDING X_p^m SPACES INTO ℓ_r^n

Gideon SCHECHTMAN
Department of Theoretical Mathematics
The Weizmann Institute of Science
Rehovot 76100, ISRAEL

§0 Introduction

For $2 < q < \infty$ and $w = \{w_i\}_{i=1}^m$ with $0 < w_i \le 1$, $i = 1, \dots, m$, and $\sum_{i=1}^m w_i^{2q/(q-2)} > 1$ let $X_{q,w}^m$ be $\mathbb{R}^m$ with the norm

$$\|\{a_i\}_{i=1}^m\| = max\left\{\left(\sum_{i=1}^m |a_i|^q\right)^{1/q}, \left(\sum_{i=1}^m w_i^2 a_i^2\right)^{1/2}\right\}.$$

Rosenthal [Ro1] proved that these spaces are K-isomorphic to K-complemented subspaces of L_q, where K depends only on q. In particular their duals, which we shall denote by $X_{p,w}^m = (X_{q,w}^m)^*$, $\frac{1}{p} + \frac{1}{q} = 1$, are uniformly isomorphic to subspaces of L_p and thus of L_1. The spaces $X_{p,w}^m$, $1 < p < 2$, or at least their infinite dimensional analogs, are known to be large spaces; they isomorphically contain all modular (and in particular Orlicz) spaces which embed into L_p (see [Ro2], [Wo] and the remark following the proof of Theorem 9 below). In Theorem 9 below we prove that all these spaces isomorphically embed, with constant depending only on p, into ℓ_1^n, where $n \le Km$ and K depends only on p. After the first version of this paper had been written, Bourgain, Lindenstrauss and Milman proved a much more general result: Every m-dimensional subspace of L_p, $1 < p < 2$, $(1+\varepsilon)$-embeds into ℓ_1^n where $n \le C(p, \varepsilon)m$.

In §1 below we introduce a family of random variables which are going to be used in the embedding theorem. We hope that these variables, which resemble symmetric p-stable variables, are of independent interest; this is one of the reasons this paper is being written.

In §2 we prove the embedding theorem of $X_{p,w}^m$ into ℓ_1^n in the symmetric case – where all the w_i's are equal. Theorem 8 actually proves something stronger; it deals also with embeddings of subspaces of $X_{p,w}^m$ into ℓ_1^n. In §3 we prove the embedding theorem for general $X_{p,w}^m$ spaces.

Supported in part by U.S.-Israel Binational Science Foundation.

§4 deals with a somewhat different embedding problem for X_p spaces; what is the smallest n such that $X^m_{p,w}$ embeds complementably (with constants depending only on p) into ℓ^n_p. Theorem 14 gives an estimate on n. This result is of some importance in the forthcoming paper [JS2].

Notation: The notation $a \approx b$ means that for some absolute constant K, $K^{-1}b \le a \le Kb$. If some parameters appear above the equivalence sign then the constant K may depend on them but on nothing else.

§1 A class of random variables: $X(u, S, \beta)$

Fix $0 < p < 2$, $0 < \beta \le \infty$ and a sequence $S = \{s_i\}_{i=-\infty}^{\ell}$ satisfying:

(1)
$$0 < s_i < s_{i+1} \text{ for all } i\ , \quad s_i \underset{i\to-\infty}{\longrightarrow} 0 \quad \text{and} \quad s_\ell = \beta$$

(in the case $\beta = \infty$ we take $\ell = \infty$ and replace "$s_\ell = \beta$" with $s_i \underset{i\to\infty}{\longrightarrow} \infty$)

(2)
$$\frac{s_{i+1} - s_i}{s_i^{p+1}} \le \frac{1}{2} \quad \text{for all} \quad i$$

(3)
$$\frac{s_{i+1}}{s_i} \le 2 \quad \text{for all} \quad i\ .$$

Let $\{x_i\}_{i=-\infty}^{\ell} = \{x_i(S)\}_{i=-\infty}^{\ell}$ be a sequence of independent random variables with

$$P(x_i = s_i) = P(x_i = -s_i) = \frac{s_{i+1} - s_i}{2s_i^{p+1}}\ , \quad P(x_i = 0) = 1 - \frac{s_{i+1} - s_i}{s_i^{p+1}}\ .$$

Given a random variable u with $E|u|^p < \infty$, let $\{u_i\}_{i=-\infty}^{\infty}$ be independent copies of u which are also independent of the x_i's. Form

(4)
$$X(u, S, \beta) = \sum_{i=-\infty}^{\ell} x_i u_i\ .$$

We shall see shortly that this series converges in distribution and almost surely, so that $X(u, S, \beta)$ is well defined. There is some redundancy in the notation (since $\beta = s_\ell$), the reason we keep β in the notation is that some of the properties of $X(u, S, \beta)$ depend only on β and not on the particular sequence S.

PROPOSITION 1: *The series $\sum_{i=-\infty}^{\ell} x_i u_i$ converges in distribution and almost surely. The characteristic function of the sum, that is*

$$\varphi(t) = \varphi_{X(u,S,\beta)}(t) = E\ exp(it\ X(u, S, \beta)) = E\ cos(t\ X(u, S, \beta))$$

is always positive and satisfies

$$(5)\qquad \begin{aligned} \log \varphi(t) &\approx \sum_{i=-\infty}^{\ell} E\frac{\cos(us_it)-1}{s_i^{p+1}}(s_{i+1}-s_i) \\ &\overset{p}{\approx} -E\,\min\{\beta^{2-p}|ut|^2,\ |ut|^p\}\,. \end{aligned}$$

(Recall the notation; the constant in the first equivalent is absolute and in the second depends only on p.) Also, the constant in the second equivalence is bounded as long as p is bounded away from 0 and 2.

PROOF: We first evaluate the characteristic function of x_iu_i.

$$\begin{aligned} \varphi_{x_iu_i}(t) &= \frac{s_{i+1}-s_i}{s_i^{p+1}}E\cos(us_it)+1-\frac{s_{i+1}-s_i}{s_i^{p+1}} \\ &= 1+\frac{E\cos(us_it)-1}{s_i^{p+1}}(s_{i+1}-s_i)\,. \end{aligned}$$

Since

$$\left|E\frac{\cos(us_it)-1}{s_i^{p+1}}(s_{i+1}-s_i)\right| \le \frac{1}{2}$$

(by (2)), we get that (with absolute constant)

$$\log \varphi_{x_iu_i}(t) \approx E\frac{\cos(us_it)-1}{s_i^{p+1}}(s_{i+1}-s_i)\,.$$

Consequently,

$$(6)\qquad \begin{aligned} \log \prod_{i=-\infty}^{\ell} \varphi_{x_iu_i}(t) &= \sum_{i=-\infty}^{\ell} \log\left(1+E\frac{\cos(us_it)-1}{s_i^{p+1}}(s_{i+1}-s_i)\right) \\ &\approx E\sum_{i=-\infty}^{\ell} \frac{\cos(us_it)-1}{s_i^{p+1}}(s_{i+1}-s_i)\,. \end{aligned}$$

In the rest of the proof we use only property (3) of the sequence $\{s_i\}$. For any w,

$$(7)\qquad \sum_{i=-\infty}^{\ell} \frac{1-\cos ws_i}{s_i^{p+1}}(s_{i+1}-s_i) \overset{p}{\approx} \min\{\beta^{2-p}w^2, |w|^p\}$$

with constants depending only on p and which are bounded when p stays bounded away from 0 and 2.

Indeed, if $|w|\beta \le 1$ then

$$1-\cos ws_i \approx w^2s_i^2 \quad \text{for all } i$$

and, since $1 < \frac{s_{i+1}}{s_i} \leq 2$, we get in this case

$$\sum_{i=-\infty}^{\ell} \frac{1 - cos\ ws_i}{s_i^{p+1}}(s_{i+1} - s_i) \approx w^2 \sum_{i=-\infty}^{\ell} \frac{s_{i+1} - s_i}{s_i^{p-1}}$$

$$\approx w^2 \int_0^\beta \frac{1}{s^{p-1}} ds$$

$$= \frac{w^2 \beta^{2-p}}{2-p}$$

$$= \frac{1}{2-p} min\{\beta^{2-p} w^2, |w|^p\} .$$

If $|w|\beta > 1$ then

$$\sum_{i=-\infty}^{\ell} \frac{1 - cos\ ws_i}{s_i^{p+1}}(s_{i+1} - s_i) \geq \sum_{s_i \leq |w|^{-1}} \frac{1 - cos\ ws_i}{s_i^{p+1}}(s_{i+1} - s_i)$$

which, by the previous case is universally equivalent to $\frac{1}{2-p} w^2 |w|^{-(2-p)} = \frac{1}{2-p}|w|^p$. On the other hand

$$\sum_{i=-\infty}^{\ell} \frac{1 - cos\ ws_i}{s_i^{p+1}}(s_{i+1} - s_i) \leq \left(\sum_{s_i \leq |w|^{-1}} + \sum_{s_i > |w|^{-1}} \right) \frac{1 - cos\ ws_i}{s_i^{p+1}}(s_{i+1} - s_i) .$$

The first summand behaves like $|w|^p$ and the second is dominated by

$$\sum_{s_i > |w|^{-1}} \frac{s_{i+1} - s_i}{s_i^{p+1}} \approx \int_{|w|^{-1}}^{\ell} \frac{ds}{s^{p+1}} \leq \frac{1}{p}|w|^p .$$

Since, for $|w|\beta > 1$, $|w|^p = min\{\beta^{2-p} w^2, |w|^p\}$ we get (7).

Combining (6) and (7), we get

$$\text{(8)} \qquad log \prod_{i=-\infty}^{\ell} \varphi_{x_i u_i}(t) \overset{p}{\approx} - E\ min\{\beta^{2-p}|ut|^2, |ut|^p\} .$$

Since the right hand side in (8) is small for t close to zero, it follows that $\prod_{i=-\infty}^{\ell} \varphi_{x_i u_i}(t)$ converges for all t in some neighborhood of zero and is bounded away from zero in this neighborhood. Thus (see e.g. [Br], p.176-177), $\sum_{i=-\infty}^{\ell} x_i u_i$ converges almost surely and necessarily the sum $X(u, S, \beta)$ has characteristic function $\varphi(t)$ which, by (6) and (8) satisfy (5).

□

REMARKS: In the case $\beta = \infty$ we get

$$log\ \varphi(t) \overset{p}{\approx} - |t|^p E|u|^p$$

which shows the relation of $X(u,S,\beta)$ with symmetric p-stable random variables.

2. In [Sc1] a closely related family of random variables is considered. They are limits of $X(u,S,\beta)$ over a suitable net of sequences S. Their characteristic function has simpler form, $exp(-E\int_0^\beta \frac{1-cos\ ust}{s^{p+1}}ds)$, and they are even more closely related to p-stables. Unfortunately, they are not good enough for the purposes of this paper.

Next we want to compute the moments of $X(u,S,\beta)$. For $0<p<2,\ 0<\beta\le\infty$. Let

$$M_{p,\beta}(r)=min\{\beta^{2-p}r^2,|r|^p\}\ , \qquad -\infty<r<\infty$$

and let

$$\|u\|_{p,\beta}=inf\{\lambda>0; EM_{p,\beta}(\frac{|u|}{\lambda})\le 1\}\ .$$

This is a quasi-norm ($\|u+v\|\le K(\|u\|+\|v\|)$) and for $1\le p<2$ is, with universal constant, equivalent to a norm; the Orlicz norm given by the Orlicz function

$$M'_{p,\beta}(r)=\int_0^r min\{\beta^{2-p}s,s^{p-1}\}ds\ .$$

(Note however that $M_{p,\beta},M'_{p,\beta}$ are not normalized, for $\beta<1$, $M_{p,\beta}(1)=\beta^{2-p}$.)

PROPOSITION 2: *For* $0<r<p<2$,

$$(E|X(u,S,\beta)|^r)^{\frac{1}{r}}\overset{p,r}{\approx}\|u\|_{p,\beta}\ .$$

We first need a standard lemma,

LEMMA 3: *Let* X *be any random variable and let* $0<r<2$. *Then*

$$E|X|^r\approx\alpha_r\int_0^\infty\frac{min\{|log|E\ cos\ Xt\||,1\}}{t^{r+1}}dt \tag{9}$$

(with an absolute constant), where

$$\alpha_r=\left(\int_0^\infty\frac{1-cos\ t}{t^{r+1}}dt\right)^{-1}\ .$$

PROOF:

$$E|X|^r=\alpha_r E\int_0^\infty\frac{1-cos\ Xt}{t^{r+1}}dt$$

and for $-1\le a\le 1,\ 1-a\approx min\{|log|a||,1\}$ (with an absolute constant).

□

PROOF OF PROPOSITION 2: By Proposition 1 and Lemma 3,

$$E|X(u,S,\beta)|^r\overset{p}{\approx}\alpha_r\int_0^\infty min\{M_{p,\beta}(ut),1\}t^{-r-1}dt\ .$$

Assume $\|u\|_{p,\beta} = 1$ then

$$EM_{p,\beta}(ut) \leq 1 \Longleftrightarrow t \leq 1$$

and

$$E|X(u,S,\beta)|^r \overset{p}{\approx} \alpha_r \left[\int_0^1 EM_{p,\beta}(ut)t^{-r-1}dt + \int_1^\infty t^{-r-1}dt\right] . \tag{10}$$

The second integral is

$$\int_1^\infty t^{-r-1}dt = \frac{1}{r} .$$

We now evaluate the first, 1_A denotes the indicator function of the set A,

$$\begin{aligned}
&\int_0^1 EM_{p,\beta}(ut)t^{-r-1}dt \\
&= E\left[\int_0^{min\{|\beta u|^{-1},1\}} \beta^{2-p}u^2t^{1-r}dt + \int_{min\{|\beta u|^{-1},1\}}^1 |u|^p t^{p-r-1}dt\right] \\
&= E\left[\frac{1}{2-r} min\{\beta^{r-p}|u|^r, \beta^{2-p}u^2\} + \frac{1}{p-r}(|u|^p - min\{\beta^{r-p}|u|^r, |u|^p\})\right] \qquad (11) \\
&= E\left[\frac{1}{2-r}\beta^{2-p}u^2 1_{(|u|<\beta^{-1})} + \left(\frac{1}{2-r} - \frac{1}{p-r}\right)\beta^{r-p}|u|^r 1_{(|u|>\beta^{-1})}\right. \\
&\quad \left. + \frac{1}{p-r}|u|^p 1_{(|u|>\beta^{-1})}\right] .
\end{aligned}$$

Since the middle summand is negative we get that

$$\int_0^1 EM_{p,\beta}(ut)t^{-r-1}dt \leq K_{p,r}E\ min\{\beta^{2-p}u^2, |u|^p\} = K_{p,r}$$

and, by (10)

$$E|X(u,S,\beta)|^r \overset{p,r}{\approx} 1 .$$

□

In the next proposition we calculate $E|X(u,S,\beta)|^r$ for $\beta < \infty$ and $0 < p \leq r < 2$. This is not needed in the rest of this paper. We bring it for completeness only.

PROPOSITION 4: (a) *For* $0 < p < r < 2$

$$E|X(u,S,\beta)|^r \overset{p,r}{\approx} \|u\|_{p,\beta}^r \beta^{r-p} E\ M_{r,\beta}\left(\frac{|u|}{\|u\|_{p,\beta}}\right)$$

(b)

$$\begin{aligned}
E|X(u,S,\beta)|^p &\overset{p}{\approx} \|u\|_{p,\beta}^p E\ N_{p,\beta}\left(\frac{|u|}{\|u\|_{p,\beta}}\right) \\
&\overset{p}{\approx} \|u\|_{N_{p,\beta}} .
\end{aligned}$$

The last expression is the Orlicz "norm" with respect to the function $N_{p,\beta}(s) = min\left(\beta^{2-p}s^2, |s|^p L(\beta s)\right)$, where $L(s) = log(max(|s|,1)) + 1$.

PROOF: To prove (**a**) note that the proof of Proposition 2 until (11) carries over also in the case $0 < p < r < 2$. Also, the equality between the first, second and last terms in (11) holds (though the third term should be changed a bit in this case).

Now,

$$\beta^{r-p}|u|^r 1_{(|u|>\beta^{-1})} \geq |u|^p 1_{(|u|>\beta^{-1})}$$

so that (11) implies

$$\int_0^1 E\ M_{p,\beta}(ut)t^{-r-1}dt \overset{p,r}{\approx} E\ min\{\beta^{2-p}u^2, \beta^{r-p}|u|^r\} = \beta^{r-p}E\ M_{r,\beta}(|u|)\ .$$

The last expression is larger than

$$E\ M_{p,\beta}(|u|) = 1$$

so we get from (10) that, if $\|u\|_{p,\beta} = 1$ then,

$$E|X(u,S,\beta)|^r \overset{p,r}{\approx} \beta^{r-p}E\ M_{r,\beta}(|u|)\ .$$

Homogeneity now implies that, for all u,

$$E|X(u,S,\beta)|^r \overset{p,r}{\approx} \|u\|_{p,\beta}^r \beta^{r-p} E\ M_{r,\beta}\left(\frac{|u|}{\|u\|_{p,\beta}}\right)\ . \tag{12}$$

To prove (**b**) proceed as in the proof of Proposition 2 to get that if $\|u\|_{p,\beta} = 1$ then

$$E|X(u,S,\beta)|^p \overset{p}{\approx} \int_0^1 E\ M_{p,\beta}(ut)t^{-p-1}dt + \frac{1}{p}\ . \tag{13}$$

Now,

$$\begin{aligned}
&\int_0^1 E\ M_{p,\beta}(ut)t^{-p-1}dt \\
&= E\left[\int_0^{min\{|\beta u|^{-1},1\}} \beta^{2-p}u^2 t^{1-p}dt + \int_{min\{|\beta u|^{-1},1\}}^1 |u|^p t^{-1}dt\right] \\
&= E\left[\frac{1}{2-p}\beta^{2-p}u^2 1_{(|u|<\beta^{-1})} + \frac{1}{2-p}|u|^p 1_{(|u|>\beta^{-1})} + |u|^p log|\beta u| 1_{(|u|>\beta^{-1})}\right] \\
&\overset{p}{\approx} E\ N_{p,\beta}(|u|)\ .
\end{aligned} \tag{14}$$

Since $E\ N_{p,\beta}(|u|) \geq E\ M_{p,\beta}(|u|) = 1$, we get from (13) that, if $\|u\|_{p,\beta} = 1$ then

$$E|X(u,S,\beta)|^p \overset{p}{\approx} E\ N_{p,\beta}(|u|)\ .$$

By homogeneity, for any u,

$$(E|X(u,S,\beta)|^p)^{1/p} \overset{p}{\approx} \|u\|_{p,\beta}\left(E\ N_{p,\beta}(\frac{|u|}{\|u\|_{p,\beta}})\right)^{1/p}\ .$$

It remains to prove the last equivalence. Let

$$[u] = \|u\|_{p,\beta} \left(E\ N_{p,\beta}(\frac{|u|}{\|u\|_{p,\beta}}) \right)^{1/p} .$$

Since $N_{p,\beta}(|s|^{1/p})$ is equivalent to a convex function

$$K_p[u]^p \geq E\ N_{p,\beta}(|u|)$$

so that $K_p[u] \geq \|u\|_{N_{p,\beta}}$.

To prove the other side inequality assume $\|u\|_{N_{p,\beta}} = 1$ then $\|u\|_{p,\beta} \leq 1$ and, putting $\|u\| = \|u\|_{p,\beta}$

$$\begin{aligned}[u] &\stackrel{p}{\approx} \|u\| \left(E \left[\beta^{2-p} (\frac{u}{\|u\|})^2 1_{(\frac{|u|}{\|u\|} < \beta^{-1})} + (\frac{|u|}{\|u\|})^p 1_{(\frac{|u|}{\|u\|} > \beta^{-1})} \right] \right)^{1/p} \\ &+ \left(E|u|^p log \left| \frac{\beta u}{\|u\|} \right| 1_{(\frac{|u|}{\|u\|} > \beta^{-1})} \right)^{1/p} \\ &= I + II .\end{aligned}$$

By the definition of $\|u\| = \|u\|_{p,\beta}$, the first term, I, is smaller than one. Now

$$\begin{aligned} II^p &= E|u|^p log|\beta u| 1_{(\frac{|u|}{\|u\|} > \beta^{-1})} + E|u|^p 1_{(\frac{|u|}{\|u\|} > \beta^{-1})} log \frac{1}{\|u\|} \\ &\leq E|u|^p log|\beta u| 1_{(|u| > \beta^{-1})} + \|u\|^p log \frac{1}{\|u\|} .\end{aligned}$$

Both terms are bounded, the first because $\|u\|_{N_{p,\beta}} \leq 1$ and the second because $\|u\| \leq 1$. Consequently

$$[u] \leq K_p \|u\|_{N_{p,\beta}} .$$

□

In the next proposition we evaluate the L_r-norm of linear combinations of independent copies of variables of the form $X(u, S, \beta)$.

PROPOSITION 5: *Let $0 < r < p < 2$ and S, β satisfy (1), (2), (3). Let $u_1, \ldots, u_m$ be random variables with p-th moment and $a_1, \ldots, a_m \in I\!R$. Put*

$$v = m^{1/p} \sum_{j=1}^{m} 1_{(\xi = j)} a_j u_j$$

where $P(\xi = j) = \frac{1}{m}$, $j = 1, \ldots, m$, and ξ is independent of the u_j's. Let $X_j \stackrel{dist}{=} X(u_j, S, \beta)$ be independent. Then

$$\left(E \left| \sum_{j=1}^{m} a_j X_j \right|^r \right)^{1/r} \stackrel{p,r}{\approx} \|v\|_{p, \beta m^{-1/p}} .$$

PROOF: For $1 \le j \le m$, $-\infty < i \le \ell$, let $x_{i,j}$, $u_{i,j}$ be independent random variables with $x_{i,j} \overset{dist}{=} x_i$, $u_{i,j} \overset{dist}{=} u_j$ and let

$$X_j = \sum_{i=-\infty}^{\ell} x_{ij} u_{ij} \ , \qquad j = 1, \dots, m \, .$$

We now evaluate the characteristic function $\varphi_{\Sigma a_j X_j}(t) = E \; cos(t\Sigma_{j=1}^m a_j X_j)$.

$$\begin{aligned}
log \; \varphi_{\Sigma a_j X_j}(t) &= \sum_{j=1}^{m} log \; \varphi_{X_j}(a_j t) \\
&\overset{p}{\approx} \sum_{j=1}^{m} \sum_{i=-\infty}^{\ell} E \frac{cos(u_j s_i a_j t) - 1}{s_i^{p+1}} (s_{i+1} - s_i) \qquad \text{(Prop.1)} \\
&= \sum_{i=-\infty}^{\ell} \frac{1}{m} \sum_{j=1}^{m} E \frac{cos(a_j u_j s_i t) - 1}{(s_i/m^{1/p})^{p+1}} \left(\frac{s_{i+1}}{m^{1/p}} - \frac{s_i}{m^{1/p}} \right) \\
&= \sum_{i=-\infty}^{\ell} E \frac{cos(v s_i m^{-1/p} t) - 1}{(s_i m^{-1/p})^{p+1}} \left(\frac{s_{i+1}}{m^{1/p}} - \frac{s_i}{m^{1/p}} \right) .
\end{aligned}$$

We now use the proof of Proposition 1 (from (7) on) for the sequence $\{\frac{s_i}{m^{1/p}}\}$. Note that in this part of the proof of Proposition 1 only property (3) of the sequence is used. This property holds also for $\{\frac{s_i}{m^{1/p}}\}$ and we conclude

$$log \; \varphi_{\Sigma a_j X_j}(t) \overset{p}{\approx} - E \; min \left\{ (\frac{\beta}{m^{1/p}})^{2-p} |ut|^2, |ut|^p \right\} .$$

Now, as in the proof of Proposition 2,

$$\left\| \sum_{j=1}^{m} a_j X_j \right\| \overset{p,r}{\approx} \|v\|_{p, \beta m^{-1/p}} \, .$$

□

The definition of the variables $X(u, S, \beta)$ can be easily extended, modulo convergence problems, to include the case of vector valued u.

In our application the vector valued random variables we shall consider will take values in a finite dimensional space so that no problem of convergence occurs.

Let $(e_1, \dots, e_n)$ be the unit vector basis in ℓ_r^n, $0 < r < 2$. Let U have the uniform distribution over $e_1, \dots, e_n$, i.e., $P(U = e_i) = \frac{1}{n}$, $i = 1, \dots, n$, and form

$$Y = X(U, S, \beta) = \sum_{i=-\infty}^{\ell} U_i x_i$$

(where U_i are independent (also of $\{x_i\}_{i=-\infty}^{\ell}$) copies of U). Let Y_k, $k = 1, \dots, m$, be independent copies of Y.

PROPOSITION 6: *Let $0 < r < p < 2$, and let m, n be positive integers such that $n^{1/p}S$, $n^{1/p}\beta$ satisfy (1), (2), (3). Let $u = \Sigma_{k=1}^m a_k m^{1/p} 1_{A_k}$ where A_k are disjoint subsets of $[0,1]$ with $\mu(A_k) = \frac{1}{m}$. Then*

$$E\left\|\sum_{k=1}^m a_k Y_k\right\|_r^r \overset{p,r}{\approx} n^{1-\frac{r}{p}} \|u\|_{p,\bar{\beta}}^r$$

where $\bar{\beta} = (\frac{n}{m})^{1/p}\beta$.

PROOF: Let $\{U_{i,k}\}_{i=-\infty,\ k=1}^{\ell,\ m}$ be independent copies of U and $\{x_{i,k}\}_{k=1}^m$ independent copies of x_i. Then

$$\begin{aligned} E\left\|\sum_{k=1}^m a_k Y_k\right\|_r^r &= E\sum_{\ell=1}^n \left|\sum_{k=1}^m a_k \sum_{i=-\infty}^{\ell} 1_{(U_{i,k}=e_\ell)} x_{i,k}\right|^r \\ &= n\, E\left|\sum_{k=1}^m a_k X_k\right|^r \end{aligned}$$

where X_k are independent copies of $X(u, S, \beta)$ with $P(u=1) = \frac{1}{n}$, $P(u=0) = 1 - \frac{1}{n}$. X_k have the same distribution as $n^{-1/p}X(1, n^{1/p}S, n^{1/p}\beta)$. Indeed,

$$\begin{aligned} \varphi_{X(u,S,\beta)}(t) &= \prod_{i=-\infty}^{\ell}\left[E\frac{cos(s_i ut) - 1}{s_i^{p+1}}(s_{i+1} - s_i) + 1\right] \\ &= \prod_{i=-\infty}^{\ell}\left[\frac{1}{n}\frac{cos(s_i t) - 1}{s_i^{p+1}}(s_{i+1} - s_i) + 1\right] \\ &= \prod_{i=-\infty}^{\ell}\frac{cos(s_i t) - 1}{(n^{1/p}s_i)^{p+1}}(n^{1/p}s_{i+1} - n^{1/p}s_i) \\ &= \varphi_{n^{-1/p}X(1,n^{1/p}S,n^{1/p}\beta)}(t)\ . \end{aligned}$$

Consequently,

$$\begin{aligned} E\left\|\sum_{k=1}^m a_k Y_k\right\|_r^r &= n^{1-r/p} E\left|\sum_{k=1}^m a_k X_k(1, n^{1/p}S, n^{1/p}\beta)\right|^r \\ &\overset{p,r}{\approx} n^{1-r/p}\|u\|_{p,(\frac{n}{m})^{1/p}\beta}^r \qquad \text{(Prop.5)}\ . \end{aligned}$$

□

REMARK: Using Proposition 4, one can state and prove a similar statement in the case $0 < p \leq r < 2$.

§2 Embedding symmetric X_p^m spaces in ℓ_1^n

Fix a constant $\alpha > 0$ such that the sequence $S = \{s_i\}_{i=-\infty}^{-1}$, $s_i = \frac{\alpha}{(-i)^{1/p}}$, satisfies (2) (for $p \geq 1$, $\alpha = 4$ is okay). (3) is also always satisfied. For β of the form $\frac{\alpha}{\ell^{1/p}}$, $\ell \in \mathbb{N}$, and $S = \{s_i\}_{i=-\infty}^{-\ell}$ (1) is also satisfied. Fix $m, n \in \mathbb{N}$ and put, as in §1

$$Y = X(U, S, \beta) .$$

Let Y_k, $k = 1, \ldots, m$, be independent copies of Y. We already know (Proposition 6) that, for $1 < p < 2$,

$$E\left\|\sum_{k=1}^{m} a_k Y_k\right\|_{\ell_1} \overset{p}{\approx} n^{1/q}\|u\|_{p,\bar{\beta}}$$

($u = \Sigma_{k=1}^m a_k m^{1/p} 1_{A_k}$, $\frac{1}{p} + \frac{1}{q} = 1$, $\bar{\beta} = (\frac{n}{m})^{1/p}\beta$).

We now want to evaluate the probability of deviation of $\|\Sigma_{k=1}^m a_k Y_k\|_{\ell_1}$ from its expectation.

PROPOSITION 7: *Let $1 < p < \infty$ and let $u = \Sigma_{k=1}^m a_k m^{1/p} 1_{A_k}$, with A_k disjoint, $\mu(A_k) = \frac{1}{m}$ and $\|u\|_{p,\bar{\beta}} = 1$. Then, for some δ, depending only on p, and all $0 < \varepsilon < 1$,*

$$P\left(\left|\|\sum_{k=1}^{m} a_k Y_k\|_{\ell_1} - E\|\sum_{k=1}^{m} a_k Y_k\|_{\ell_1}\right| > \varepsilon n^{1/q}\right) \leq 4e^{-\delta\varepsilon^q n} .$$

PROOF: Let $\{x_i\}_{i=-\infty}^{-\ell}$ be the sequence corresponding to S, and let $\left\{\{x_{i,j}\}_{i=-\infty}^{-\ell} U_{i,j}\right\}_{j=1}^{m}$ be independent copies of $\{x_i\}_{i=-\infty}^{-\ell}, U_i$. Then

$$\left\|\sum_{k=1}^{m} a_k Y_k\right\| \overset{dist}{=} \left\|\sum_{k=1}^{m}\sum_{i=-\infty}^{-\ell} a_k U_{i,k} x_{i,k}\right\| .$$

Arrange $\{i, k\}$ in any linear order, $\leq$, and let $\mathcal{F}_{i,k}$ be the σ-field generated by $\{U_{t,j}\ x_{t,j}\}_{(t,j)\leq(i,k)}$. Let $E_{i,k}$ be the corresponding conditional expectation and let $d_{i,k}$ be the martingale difference sequence obtained from the function

$$\left\|\sum_{k=1}^{m}\sum_{i=-\infty}^{-\ell} a_k U_{i,k} x_{i,k}\right\|$$

using these conditional expectations. Then easily (see e.g. [MS] bottom of p.95; this is due to Yurinskii),

$$|d_{i,k}| \leq |a_k x_{i,k}| + E|a_k x_{i,k}| .$$

In particular,

$$(15) \qquad |d_{i,k}| \leq 2|a_k|\alpha|i|^{-1/p}$$

and, for any subset A of $\{1,\dots,m\}$,

(16) $$M_A = \sup_{\substack{k\in A\\ i}} \|d_{i,k}\|_\infty \le 2\beta \max_{k\in A} |a_k|$$

(17) $$\left\| \{\|d_{i,k}\|_\infty\}_{i=-\infty,k\in A}^{-\ell} \right\|_{p,\infty} \le 2\alpha \left(\sum_{k\in A} |a_k|^p \right)^{1/p}$$

(where $\|\{\lambda_t\}_{t=1}^\infty\|_{p,\infty} = \sup_t\ \lambda_t^*\, t^{1/p}$, $\{\lambda_t^*\}_{t=1}^\infty$ is the decreasing rearrangement of $\{|\lambda_t|\}_{t=1}^\infty$)

(18) $$\sigma_A^2 = \sum_{\substack{k\in A\\ i}} E_{(i,k)'}(d_{i,k})^2 \le K_p \beta^{2-p} \sum_{k\in A} a_k^2$$

where K_p depends only on p and $(i,k)'$ is the immediate predecessor of (i,k). Let us prove (18).

$$\begin{aligned}
\sigma_A^2 &\le 2 \sum_{k\in A} a_k^2 \sum_{i=-\infty}^{-\ell} E_{(i,k)'}(x_{i,k})^2 \\
&= 2 \sum_{k\in A} a_k^2 \sum_{i=-\infty}^{-\ell} E(x_{i,k})^2 \\
&= 2 \sum_{k\in A} a_k^2 \sum_{i=-\infty}^{-\ell} \frac{\alpha^2}{|i|^{2/p}} (|i|-1)^{-1/p} - |i|^{-1/p})|i|^{1+1/p}\alpha^{-p} \\
&\le 4\ \alpha^{2-p} \sum_{i=\ell}^{\infty} \frac{1}{i^{2/p}} \sum_{k\in A} a_k^2 \\
&\le K_p \alpha^{2-p} \ell^{1-2/p} \sum_{k\in A} a_k^2 \\
&\le K_p \beta^{2-p} \sum_{k\in A} a_k^2\ .
\end{aligned}$$

Let now $u = \Sigma_{k=1}^m a_k m^{1/p} 1_{A_k}$ and assume without loss of generality $\|u\|_{p,\bar\beta} = 1$. Put

$$A = \{1 \le k \le m;\ |a_k| < \bar\beta^{-1} m^{-1/p}\}\ , \qquad B = \{1,\dots,m\}\backslash A$$

then

(19) $$\beta^{2-p} n^{2/p-1} \sum_{k\in A} a_k^2 + \sum_{k\in B} |a_k|^p = 1\ .$$

$$\begin{aligned}
&P\left(\left| \|\sum a_k Y_k\| - E\|\sum a_k Y_k\| \right| > \varepsilon n^{1/q} \right) = P\left(\left| \sum_{i,k} d_{i,k} \right| > \varepsilon n^{1/q} \right) \\
&\le P\left(\left| \sum_{\substack{k\in A\\ i}} d_{i,k} \right| > \frac{\varepsilon n^{1/q}}{2} \right) + P\left(\left| \sum_{\substack{k\in B\\ i}} d_{i,k} \right| > \frac{\varepsilon n^{1/q}}{2} \right) \\
&= I + II\ .
\end{aligned}$$

By the martingale version of Prokhorov's inequality (Proposition 3.1 in [JSZ]),

$$\begin{aligned} I &\le 2\ exp\left(\frac{-\varepsilon n^{1/q}}{2M_A} arc\ sinh\ \frac{\varepsilon n^{1/q} M_A}{2\sigma_A^2}\right) \\ &\le 2\ exp(-\delta\varepsilon^2 n) \qquad (\text{using} 16, 18, 19) \end{aligned}$$

where δ depends only on p (note that for small ε, $arc\ sinh\ \varepsilon \approx \varepsilon$).

By a martingale inequality due to Pisier (see Lemma 8.4 in [MS]),

$$\begin{aligned} II &\le 2\ exp\left(-\delta\varepsilon^q n / \left\| \{\|d_{i,k}\|_\infty\}_{i=-\infty}^{-\ell}{}_{k\in B} \right\|_{p,\infty}^q \right) \\ &\le 2\ exp(-\delta'\varepsilon^q n) \qquad (\text{using} 17)\ , \end{aligned}$$

again, with δ depending only on p.

□

Let us recall now the definition of the X_p spaces. For $w = \{w_i\}_{i=1}^m$ $0 < w_i \le 1$ let $X_{p,w} = X_{p,w}^m$ be $\mathbb{R}^m$ with the modular space norm given by the functions $M_{p,w_i^{(p-2)/2}}$ (we shall denote this norm $|||\cdot|||_{p,w}$ to distinguish it from the function space norm)

$$|||\{a_i\}_{i=1}^m|||_{p,w} = \inf\left\{\lambda > 0; \sum_{i=1}^m min\{(\frac{a_i}{w_i\lambda})^2, |\frac{a_i}{\lambda}|^p\} \le 1\right\}\ .$$

This is a quasi-norm which, for $1 \le p \le 2$, is equivalent to a norm. These spaces were studied by Rosenthal [Ro1, Ro2] who proved among other things that, for $1 < p < 2$, they are uniformly isomorphic to uniformly complemented subspaces of L_p. Note that if $\Sigma_{i=1}^m w_i^{\frac{2p}{2-p}} \le 1$ then $|||\cdot|||_{p,w}$ is universally equivalent to the ℓ_p norm so we shall always assume $\Sigma_{i=1}^m w_i^{\frac{2p}{2-p}} > 1$. Note also that for $w_i = (m^{1/p}\beta)^{\frac{p-2}{2}}$, $i = 1, \dots, m$

$$|||(a_i)|||_{p,w} = \|\sum a_i m^{1/p} 1_{[\frac{i-1}{m}, \frac{i}{m}]}\|_{p,\beta}\ .$$

The following is the main theorem of this section.

THEOREM 8: *For any $1 < p < 2$ there exists a constant $0 < K < \infty$, depending only on p, such that if $w = \{w\}_{i=1}^m$ is any constant sequence with $0 < w \le 1$ and $mw^{\frac{2p}{2-p}} \ge 1$ and if $n \le m$ is any positive integer such that $nw^{\frac{2p}{2-p}} \ge 1$ then any n-dimensional subspace of $X_{p,w}^m$ K-embeds into ℓ_1^{Kn}.*

PROOF: Let $\bar\beta = w^{\frac{2}{p-2}} m^{-1/p}$, $\beta = (\frac{m}{n})^{1/p}\bar\beta = w^{\frac{2}{p-2}} n^{-1/p} \le 1$. Then $w = (m^{1/p}\bar\beta)^{\frac{p-2}{2}}$ so that

$$|||(a_i)|||_{p,w} = \|\sum a_i m^{1/p} 1_{[\frac{i-1}{m}, \frac{i}{m}]}\|_{p,\bar\beta}\ .$$

By Propositions 6 and 7 (one may assume without loss of generality that $\beta = \frac{\alpha}{\ell^{1/p}}$ for some positive integer ℓ) there is a family of linear operators from $X_{p,w}^m$ into ℓ_1^{Kn}, T_t, t ranges over some probability space, such that for all $\bar a \in X_{p,w}$ $E\|T_t\bar a\| \overset{p}{\approx} (Kn)^{1/q}|||\bar a|||_{p,w}$ and

$$P\{t; \|T_t\bar a\| \overset{p}{\approx} E\|T_t\bar a\|\} \ge 1 - 4e^{-\delta K n}\ .$$

The rest of the proof is now a standard, approximation by ε-net, argument.

□

REMARK: The random embedding we use here is very similar to the one Pisier used in [Pi]. Indeed the Y_k's we are using here have the form $Y_k \overset{dist}{=} \Sigma_{j=\ell}^{\infty} \frac{t_j}{j^{1/p}} U_j$ where t_j are independent three valued symmetric random variables with $P(t_j \neq 0)$ bound away from zero, while the corresponding $\tilde{S}_i$'s Pisier is using are of the form $\sum_{j=1}^{\infty} \frac{r_j}{j^{1/p}} U_j$, r_j being the Rademacher functions.

§3 Embedding non symmetric X_p^m spaces in ℓ_1^n

Here we prove

THEOREM 9: *For any $1 < p < 2$ there exists a constant $0 < K < \infty$ depending only on p such that for any $k, n \in I\!N$ with $Kk < n$ and any $\boldsymbol{w} = \{w_i\}_{i=1}^k$, $0 < w_i \leq 1$, $X_{p,\boldsymbol{w}}^k$ K-embeds into ℓ_1^n.*

The theorem will follow from Theorem 8 together with the following proposition which gives a (deterministic, rather than probabilistic) embedding of general X_p space in a symmetric one (that is, one for which all the w's are equal). The proof is basically contained in [Ro1].

PROPOSITION 10: *Let k be a positive integer and let $\boldsymbol{w} = \{w_i\}_{i=1}^k$ where $k^{\frac{p-2}{2p}} < w_i \leq 1$, $i = 1, \ldots, k$. Then $X_{p,\boldsymbol{w}}^k$ embeds with an absolute constant into $X_{p,\boldsymbol{u}}^m$ where $\boldsymbol{u} = \{u_i\}_{i=1}^m$ $u_i = u = k^{(p-2)/2p}$ and $k\Sigma_{i=1}^k w_i^{2p/(2-p)} \leq m \leq 2k\Sigma_{i=1}^k w_i^{2p/(2-p)}$.*

PROOF: Let $\{e_i\}_{i=1}^m$ be the unit vector basis in $X_{p,\boldsymbol{u}}^m$. If $\sigma \subseteq \{1, \ldots, m\}$ with $|\sigma| \leq k$ then $|||\sum_{i\in\sigma} e_i|||_{p,\boldsymbol{u}} = |\sigma|^{1/p}$. Indeed,

$$\sum_{i\in\sigma} min\left\{(\frac{1}{u\lambda})^2, |\frac{1}{\lambda}|^p\right\} = min\left\{\frac{|\sigma|}{(u\lambda)^2}, \frac{|\sigma|}{|\lambda|^p}\right\}$$

and for $\lambda = |\sigma|^{1/p}$ we get

$$= min\left\{\frac{|\sigma|^{1-2/p}}{u^2}, 1\right\} = 1$$

(since $|\sigma|^{1-2/p} \leq k^{1-2/p} = u^2$).

Let now $\sigma_1, \sigma_2, \ldots, \sigma_k$ be disjoint subsets of $\{1, \ldots, m\}$ with

$$|\sigma_j| = \ell_j \leq k\ .$$

Then

$$|||\sum_{j=1}^k a_j \ell_j^{-1/p} \sum_{i\in\sigma_j} e_i|||_{p,\boldsymbol{u}} = 1$$

if and only if

$$1 = \sum_{j=1}^{k} \ell_j \, min\left\{ (\frac{a_j}{u\ell_j^{1/p}})^2, \frac{|a_j|^p}{\ell_j} \right\}$$

$$= \sum_{j=1}^{k} min\left\{ (u\ell_j^{(2-p)/2p})^{-2}, |a_j|^p \right\} .$$

Putting $w_i = u\ell_i^{(2-p)/2p}$, we get

$$|||\{a_j\}_{j=1}^k|||_{p,w} = ||| \sum_{j=1}^{k} a_j \ell_j^{-1/p} \sum_{i\in\sigma_j} e_i |||_{p,u} \, .$$

If we choose $\ell_i = w_i^{2p/(2-p)}k$ (assuming this is an integer) then $\ell_i \le k$, $w_i = u\ell_i^{(2-p)/2p}$, and the proposition follows. Note that in the case that $w_i^{2p/(2-p)}k$ are integers we actually get an isometric embedding. In the general case a simple modification of the proof above will yield an absolute constant.

□

REMARK: Note that the image of $X_{p,w}^k$ in $X_{p,u}^m$ is complemented with a projection of norm one.

PROOF OF THEOREM 9: Note first that we may assume $w_i > k^{(p-2)/2p}$. Indeed $[e_i]_{w_i \le k^{(p-2)/2p}}$ span an ℓ_p space of dimension smaller than or equal to k and it embeds nicely into ℓ_1^n by [JS]. Let $m \le 2k\Sigma_{i=1}^k w_i^{2p/(2-p)}$ and u be as in Proposition 10. Then $X_{p,w}^k$ can be viewed as a subspace of $X_{p,u}^m$ of dimension k with $ku^{2p/(2-p)} = 1$. Then, by Theorem 8, $X_{p,w}^k$ K-embeds into ℓ_1^n.

□

§4 Embedding X_p^m complementably into ℓ_p^n

For $2 < q < \infty$, $0 < w_i \le 1$ and $w = \{w_i\}_{i=1}^m$, $X_{q,w}^m$ is $\mathbb{R}^m$ with the norm

$$|||\{a_i\}|||_{q,w} = max\left\{ (\sum_{i=1}^{m} |a_i|^q)^{1/q}, (\sum_{i=1}^{m} (w_i\, a_i)^2)^{1/2} \right\} .$$

This is the dual to $X_{p,w}^m$, $\frac{1}{p} + \frac{1}{q} = 1$. As we remarked above, these spaces embed uniformly into L_q and (under a suitable embedding) are uniformly complemented. In this section we shall show (Theorems 13 and 14 below) that one can actually embed $X_{p,w}^m$ $(X_{q,w}^m)$ into ℓ_p^n (ℓ_q^n), where n is small with respect to m, with constant depending only on p and projection of norm depending only on p.

We first need a specific complemented embedding of $X_{p,w}^m$ into L_p in the symmetric case $w_i = w$.

PROPOSITION 11: *Let $\{A_i\}_{i=1}^m$ be an exchangeable sequence of sets in $[0,1]$ (i.e. the distribution of $(A_{\pi(1)},\dots,A_{\pi(m)})$ is the same for all permutation π of $\{1,\dots,m\}$). Assume $\mu(A_i)=\mu=\frac{k}{m}$ for some positive integer k and that each point t in $[0,1]$ belongs to exactly $k=\mu m$ of the sets A_i. Put $x_i=\mu^{-1/q}1_{A_i}\otimes r_i$, $2<q<\infty$, (i.e. $x_i(s,t)=\mu^{-1/q}1_{A_i}(s)r_i(t)$) where r_i are the Rademacher functions. Then*

$$\left\|\sum_{i=1}^m a_ix_i\right\|_q \overset{q}{\approx} max\left\{(\sum_{i=1}^m |a_i|^q)^{1/q}, \mu^{\frac{1}{2}-\frac{1}{q}}(\sum_{i=1}^m a_i^2)^{\frac{1}{2}}\right\} \tag{20}$$

and the orthogonal projection onto $[x_i]_{i=1}^m$ has norm which depends only on q.

PROOF: The equivalence (20) is proved in [JMST, Th.1.1]. Note that $(\Sigma x_i^2)^{\frac{1}{2}}\equiv\sqrt{\mu m}$. The proof of the complementation is similar to the one in [Ro1]: Let $y_i=\mu^{-1/p}1_{A_i}\otimes r_i$, $\frac{1}{p}+\frac{1}{q}=1$, then the orthogonal projection is

$$Pf=\sum_{i=1}^m(\int y_if)x_i\ .$$

Now,

$$\begin{aligned}\left(\sum_{i=1}^m|\int y_if|^q\right)^{1/q} &= \sup_{(\Sigma_{i=1}^m|a_i|^p)^{1/p}}\int\left(\sum_{i=1}^m a_iy_i\right)f\\ &\le \sup_{(\Sigma_{i=1}^m|a_i|^p)^{1/p}}\left\|\sum_{i=1}^m a_iy_i\right\|_p\|f\|_q\\ &\le\|f\|_q\end{aligned}\tag{21}$$

since L_p is of type p.

Also,

$$\begin{aligned}\mu^{\frac{1}{2}-\frac{1}{q}}\left(\sum_{i=1}^m(\int y_if)^2\right)^{1/2} &= \left(\sum_{i=1}^m(\int\frac{1_{A_i}\otimes r_i}{\mu^{1/2}}f)^2\right)^{1/2}\\ &\le\|f\|_2\\ &\le\|f\|_q\ .\end{aligned}\tag{22}$$

It thus follows from (20), (21), (22) that $\|P\|$ depends only on q.

□

REMARKS: 1. For each $k\le m$ there is a sequence of sets $A_1,\dots,A_m$ with the above property: Partition $[0,1]$ into $\binom{m}{k}$ sets of equal measure $\{B_K\}$ where K ranges over all subsets of $\{1,\dots,m\}$ of cardinality k and put $A_i=\bigcup_{i\in K}B_K$.

2. The reason we consider the particular embedding of $X_{q,w}^m$, given in Proposition 11, rather than say the one given in [Ro1] is that we need a good bound on the L_∞ norm of the functions in $[x_i]_{i=1}^m$ and $[y_i]_{i=1}^m$.

LEMMA 12: *With the notations of Proposition 11 and its proof*

(23) $$\left\|\sum_{i=1}^{m} a_i x_i\right\|_\infty \leq K_q \, min\{m^{1/2}, m^{1/q}\mu^{1/p-1/q}\} \left\|\sum_{i=1}^{m} a_i x_i\right\|_q$$

(24) $$\left\|\sum_{i=1}^{m} a_i y_i\right\|_\infty \leq K_q \mu^{1/2-1/p} m^{1/2} \left\|\sum_{i=1}^{m} a_i y_i\right\|_p$$

(25) $$\left\|\sum_{i=1}^{m} a_i x_i\right\|_\infty \leq m^{1/2} \left\|\sum_{i=1}^{m} a_i x_i\right\|_2$$

and

(26) $$\left\|\left(\sum_{i=1}^{m} a_i x_i\right) \sum_{j=1}^{m} |a_j|^{q-1}(sign\ a_j) y_j\right\|_\infty \leq m \left(\sum_{i=1}^{m} |a_i|^q\right)^{1/q}$$

for all $a_1, \ldots, a_m$.

PROOF: Let $\{a_i^*\}_{i=1}^m$ be the decreasing rearrangement of $\{|a_i|\}_{i=1}^m$. Then, if $\|\Sigma_{i=1}^m a_i x_i\|_q = 1$,

$$\begin{aligned}\left\|\sum_{i=1}^{m} a_i x_i\right\|_\infty &= \frac{1}{\mu^{1/q}} \sum_{i=1}^{\mu m} a_i^* \\ &< \mu^{1/2-1/q} m^{1/2} \left(\sum_{i=1}^{m} a_i^2\right)^{1/2} \\ &\leq K_q \sqrt{m} \, .\end{aligned}$$

Also,

$$\begin{aligned}\left\|\sum_{i=1}^{m} a_i x_i\right\|_\infty &= \frac{1}{\mu^{1/q}} \sum_{i-1}^{\mu m} a_i^* \\ &\leq m^{1/p} \mu^{1/p-1/q} \left(\sum_{i=1}^{m} |a_i|^q\right)^{1/q} \\ &\leq K_q m^{1/p} \mu^{1/p-1/q} \, .\end{aligned}$$

This proves (23). Inequalities (24) and (25) are proved similarly. To prove (26) assume $\Sigma_{i=1}^m |a_i|^q = 1$. Then, pointwise,

$$\begin{aligned}\left|\left(\sum_{i=1}^{m} a_i x_i\right) \sum_{j=1}^{m} |a_j|^{q-1}(sign\ a_j) y_j\right| &\leq \frac{1}{\mu^{1/q}} \sum_{i=1}^{\mu m} a_i^* \frac{1}{\mu^{1/p}} \sum_{j=1}^{\mu m} (a_j^*)^{q-1} \\ &\leq \frac{1}{\mu} (\mu m)^{1/p} \left(\sum_{i=1}^{m} |a_i|^q\right)^{1/q} (\mu m)^{1/q} \left(\sum_{j=1}^{m} |a_j|^q\right)^{1/p} \\ &= m \, .\end{aligned}$$

□

We are now ready to prove

THEOREM 13: *Let* $2 < q < \infty$ *and let* $m^{(2-q)/2q} < w \le 1$. *Let* $\mathbf{w} = \{w\}_{i=1}^m$ *then* $X_{q,\mathbf{w}}^m$ *K-embeds as a K-complemented subspace into* ℓ_q^n *provided*

$$n \ge K\ min\{m^{1+q/2}, m^q w^{2q}\}\ ,$$

where K depends on q only.

REMARK: If $w \le m^{(2-q)/2q}$ then $X_{q,\mathbf{w}}^m \overset{q}{\approx} \ell_q^m$.

PROOF: Let $\mu = w^{2q/(q-2)}$. We use Proposition 4 of [Sc2] and its proof to get, in L_q^n (L_q on the measure space $\{1, \ldots, n\}$ with the uniform measure, we shall say shortly what n is), a sequence $\{\bar{z}_i\}_{i=1}^m$ of functions such that $|\bar{z}_i|$ are indicator functions and, putting $\bar{x}_i = \frac{1}{\mu^{1/q}}\bar{z}_i$, $\bar{y} = \frac{1}{\mu^{1/p}}\bar{z}_i$, we have

$$\frac{1}{2}\left\|\sum_{i=1}^m a_i x_i\right\|_q \le \left\|\sum_{i=1}^m a_i \bar{x}_i\right\|_q \le 2\left\|\sum_{i=1}^m a_i x_i\right\|_q \tag{27}$$

$$\frac{1}{2}\left\|\sum_{i=1}^m a_i y_i\right\|_p \le \left\|\sum_{i=1}^m a_i \bar{y}_i\right\|_p \le 2\left\|\sum_{i=1}^m a_i y_i\right\|_p \tag{28}$$

$$\frac{1}{2}\left(\sum_{i=1}^m a_i^2\right)^{1/2} \le \left\|\sum_{i=1}^m \frac{a_i \bar{z}_i}{\mu^{1/2}}\right\|_2 \le 2\left(\sum_{i=1}^m a_i^2\right)^{1/2} \tag{29}$$

for all $\{a_i\}_{i=1}^m$, and

$$\frac{1}{2}\sum_{i=1}^m |a_i|^q \le \int\left(\sum_{i=1}^m a_i \bar{x}_i\right)\sum_{j=1}^m |a_j|^{q-1}(sign\ a_j)\bar{y}_j \le 2\sum_{i=1}^m |a_i|^q \tag{30}$$

for all $\bar{a} = \{a_i\}_{i=1}^m$ in an ε-net in the sphere of $X_{q,\mathbf{w}}^m$ (ε will be chosen shortly).

Indeed (27), (28), (29) follow directly from (23), (24), (25) and Proposition 4 in [Sc2] and its proof provided

$$n \ge C_q min\{m^{1+q/2}, m^{1+q/p}\mu^{q/p-1}\}$$

$$n \ge C_q m^{1+p/2}\mu^{p/2-1}$$

and

$$n \ge C_q\ m^2\ .$$

The inequality (30) follows similarly, using (26) and the fact that

$$\int\left(\sum_{i=1}^m a_i x_i\right)\sum_{j=1}^m |a_j|^{q-1}(sign\ a_j)y_j = \sum_{i=1}^m |a_i|^q\ ,$$

provided

$$n \ge C\ m^2\ .$$

If $w > m^{(2-q)/2q}$ then $\mu > \frac{1}{m}$ and the inequalities for n reduce to

$$n \geq C_q \; min\{m^{1+q/2}, m^{1+q/p}\mu^{q/p-1}\} = C_q \; min\{m^{1+q/2}, m^q w^{2q}\} .$$

We claim that inequalities (27)-(30) imply that the orthogonal projection is bounded on $[\bar{x}_i]_{i=1}^m$ (the problem of course is that $\{\bar{x}_i, \bar{y}_i\}$ is not a biorthogonal system). This together with (27) will finish the proof. Define a norm $[\![\cdot]\!]$ on $\{\bar{x}_i\}_{i=1}^m$ by

$$\left[\!\left[\sum_{i=1}^m a_i\bar{x}_i\right]\!\right] = sup\left\{\left\langle \sum_{i=1}^m a_i\bar{x}_i, \sum_{i=1}^m b_j\bar{y}_j\right\rangle ; \left\|\sum_{j=1}^m b_j\bar{y}_j\right\|_p = 1\right\} .$$

We claim that this norm is equivalent to $\|\Sigma_{i=1}^m a_i\bar{x}_i\|_q$. Clearly,

$$\left[\!\left[\sum_{i=1}^m a_i\bar{x}_i\right]\!\right] \leq K_q \left\|\sum_{i=1}^m a_i\bar{x}_i\right\|_q .$$

To prove the other inequality recall that

$$\left\|\sum_{i=1}^m a_i\bar{x}_i\right\|_q \leq K_q \; max\left\{\left(\sum_{i=1}^m |a_i|^q\right)^{1/q} , \mu^{1/2-1/q}\left(\sum_{i=1}^m a_i^2\right)^{1/2}\right\} .$$

Now for $\bar{a}$ in the ε-net for which (30) holds

$$\sum_{j=1}^m |a_j|^q \leq 2\int\left(\sum_{i=1}^m a_i\bar{x}_i\right)\sum_{j=1}^m |a_j|^{q-1}(sign\; a_j)\bar{y}_j$$

and

$$\left\|\sum_{j=1}^m |a_j|^{q-1}(sign\; a_j)\bar{y}_j\right\|_p \leq 2\left(\sum_{j=1}^m |a_j|^{p(q-1)}\right)^{1/p}$$
$$\leq 2\left(\sum_{j=1}^m |a_j|^q\right)^{1/q} .$$

Consequently, if $1 = \|\bar{a}\| = (\Sigma|a_j|^q)^{1/q}$, then

$$\left\|\sum_{i=1}^m a_i\bar{x}_i\right\|_q \leq K_q\left(\sum_{i=1}^m |a_i|^q\right)^{1/q}$$
$$\leq K'_q \left[\!\left[\sum_{i=1}^m a_i\bar{x}_i\right]\!\right] .$$

On the other hand, if $1 = \|\bar{a}\| = \mu^{1/2-1/q}(\Sigma_{i=1}^m a_i^2)^{1/2}$, then

$$\int\sum_{i=1}^m a_i\bar{x}_i\sum_{j=1}^m a_j\bar{y}_j = \left\|\sum_{i=1}^m a_i\frac{\bar{z}_i}{\mu^{1/2}}\right\|_2^2$$
$$\geq \frac{1}{4}\sum_{i=1}^m a_i^2$$
$$\geq K_q^{-1}\mu^{1/q-1/2}\left(\sum_{i=1}^m a_i^2\right)^{1/2}\left\|\sum_{i=1}^m a_i\bar{x}_i\right\|_q$$

and

$$\frac{\|\Sigma_{j=1}^m a_j \bar{y}_j\|_p \mu^{1/2-1/q}}{(\Sigma_{i=1}^m a_i^2)^{1/2}} \le \frac{\|\sum_{j=1}^m a_j \bar{y}_j\|_2 \mu^{1/2-1/q}}{(\sum_{i=1}^m a_i^2)^{1/2}} = 1 .$$

So also in this case

$$\left\|\sum_{i=1}^m a_i \bar{x}_i\right\|_q \le K_q \left[\!\!\left[\sum_{i=1}^m a_i \bar{x}_i\right]\!\!\right]$$

and we conclude that for all $\bar{a} = \{a_i\}_{i=1}^m$ in an ε-net in the sphere of $X_{q,w}^m$

$$K_q^{-1} \left[\!\!\left[\sum_{i=1}^m a_i \bar{x}_i\right]\!\!\right] \le \left\|\sum_{i=1}^m a_i \bar{x}_i\right\|_q \le K_q \left[\!\!\left[\sum_{i=1}^m a_i \bar{x}_i\right]\!\!\right] . \tag{31}$$

If ε is small enough with respect to K_q we get the same for all $\bar{a}$ in $X_{q,w}^m$.

Let Q be the orthogonal projection from L_q^n onto $[\bar{x}_i]_{i=1}^m$. Then the range of Q^* is $[\bar{y}_i]_{i=1}^m$ and, by (31), for $f \in L_q^n$

$$\begin{aligned} \|Qf\| &\overset{p}{\approx} sup\{\langle \bar{y}, Qf\rangle;\ \bar{y} = \sum_{i=1}^m a_i \bar{y}_i \quad \|\bar{y}\|_p = 1\} \\ &= sup\{\langle \bar{y}, f\rangle;\ \bar{y} = \sum_{i=1}^m a_i \bar{y}_i, \quad \|\bar{y}\|_p = 1\} \\ &\le \|f\|_q . \end{aligned}$$

□

Proposition 10 now implies a similar theorem for general weights w_i.

THEOREM 14: *Let $2 < q < \infty$, k a positive integer and let $w = \{w_i\}_{i=1}^k$ $0 < w_i \le 1$, $i = 1, \dots, k$, and $\Sigma_{i=1}^k w_i^{2q/(q-2)} \ge 1$. Then $X_{q,w}^k$ K-embeds as a K-complemented subspace of ℓ_q^n provided*

$$n \ge K\ min \left\{ k^{1+q/2} \left(\sum_{i=1}^k w_i^{2q/(q-2)} \right)^{1+q/2}, k^2 \left(\sum_{i=1}^k w_i^{2q/(q-2)} \right)^q \right\} ,$$

where K depends only on q.

In particular the conclusion holds if

$$n \ge K\ k^{2+q} .$$

REMARK: If $\Sigma_{i=1}^k w_i^{2q/(q-2)} < 1$ then $X_{q,w}^k = \ell_q^k$.

PROOF: Since $X_{q,w}^k$ embeds complementably (as a subsequence) into $X_{q,\bar{w}}^k \oplus \ell_q^k$ with $\bar{w}_i > k^{(2-q)/2q}$, we may assume without loss of generality $w_i > k^{(2-q)/2q}$, $i = 1, \dots, k$.

By Proposition 10 and the remark following it, $X^k_{p,\boldsymbol{w}}$ K-embeds as a 1-complemented subspace into $X^m_{p,\boldsymbol{u}}$ where $\boldsymbol{u} = \{u\}^m_{i=1}$, $u = k^{(p-2)/2p} = k^{(2-q)/2q}$, $m \leq 2k\Sigma^k_{i=1} w_i^{2p/(2-p)} = 2k\Sigma^k_{i=1} w_i^{2q/(2-q)}$ and K depends only on q. Theorem 13 implies now that, if

$$n \geq K\, min \left\{ k^{1+q/2} \left(\sum_{i=1}^{k} w_i^{2q/(q-2)} \right)^{1+q/2}, k^2 \left(\sum_{i=1}^{k} w_i^{2q/(q-2)} \right)^{q} \right\},$$

then $X^m_{p,\boldsymbol{u}}$, and thus also $X^k_{p,\boldsymbol{w}}$, K-embed as a K-complemented subspace of ℓ^n_q, where K depends only on q.

□

REFERENCES

[Br] L. Breiman, Probability, Addison-Wesley, 1968.

[JMST] W.B. Johnson, B. Maurey, G. Schechtman and L Tzafriri, Symmetric structures in Banach spaces, Memoirs of the A.M.S., **217**, 1979.

[JS1] W.B. Johnson and G. Schechtman, Embedding ℓ_p^m into ℓ_1^n, Acta Math., **149** (1982), 71-85.

[JS2] W.B. Johnson and G. Schechtman, Sums of independent random variables in rearrangement invariant function spaces, in preparation.

[JSZ] W.B. Johnson, G. Schechtman and J. Zinn, Best constants in moment inequalities for linear combinations of independent and exchangeable random variables, Ann. Prob., **13** (1985), 234-253.

[MS] V.D. Milman and G. Schechtman, Asymptotic Theory of Finite Dimensional Normed Spaces, Lecture Notes in Math., Vol. **1200**, Springer 1986.

[Pi] G. Pisier, On the dimension of the ℓ_p^n-subspaces of Banach spaces, for $1 \leq p < 2$, Trans. A.M.S., **276** (1983), 201-211.

[Ro1] H.P. Rosenthal, On the subspaces of L^p $(p > 2)$ spanned by sequences of independent random variables, Israel J. Math, **8** (1970), 273-303.

[Ro2] H.P. Rosenthal, On the span in L^p of sequences of indepenedent random variables (II), Proc. of the 6-th Berkeley Symp. on Prob. and Stat., Berkeley, Calif., 1971.

[Sc1] G. Schechtman, Fine embeddings of finite dimensional subspaces of L_p, $1 \leq p < 2$ into finite dimensional normed spaces II, Longhorn Notes, Univ. of Texas, 1984/85.

[Sc2] G. Schechtman, More on embedding subspaces of L_p in ℓ_r^n, Compositio Math., to appear.

[Wo] J.Y.T. Woo, On a class of universal modular sequence spaces, Israel J. Math., **20** (1975), 193-215.

SOME REMARKS ON URYSOHN'S INEQUALITY AND VOLUME RATIO OF COTYPE 2-SPACES

V.D. Milman
School of Mathematical Sciences
Raymond and Beverly Sackler
Faculty of Exact Sciences
Tel Aviv University
Tel Aviv, Israel

This note consists of two parts. In the first part, we give a new proof of Urysohn's inequality relating a volume ratio of a central symmetric convex body K and the euclidean ball to the integral average E_K^*. We use this inequality in the second part through entropy estimations to give a new and very simple proof of the result from [BM1] that the volume ratios of the cotype 2-spaces are uniformly bounded by a constant depending only on the cotype 2-constant of a space.

Let $(\mathbb{R}^n, \|\cdot\|, |\cdot|)$ be a normed space $X = (\mathbb{R}^n, \|\cdot\|)$ equipped with a euclidean structure (i.e., with an inner product (x,y) and the norm $|x| = (x,x)^{1/2}$). Then the dual norm $\|x\|^* = \sup\{|(x,y)| : \|y\| \le 1\}$. Let $K = K(X) = \{x \in \mathbb{R}^n, \|x\| \le 1\}$ be the unit ball of X and $D = K\big((\mathbb{R}^n; |\cdot|)\big)$ be the euclidean ball. We also denote $S = \partial D$ the euclidean $(n-1)$-dimensional sphere and equip it with the probability rotation invariant measure μ. Clearly the unit ball $K(X^*)$ of the dual space $X^* = (\mathbb{R}^n, \|\cdot\|^*)$ is the polar body K°. We also consider the following integral characteristics of the normed space

$$E = E_K = E(X) = \int\limits_{x\in S} \|x\| d\mu(x) \quad \text{and} \quad E^* = E(X^*) \ (= E_K{}^* = E_{K^\circ})$$

and use the notions and basic properties of type and cotype of a space X (see, e.g. [MSch] for the necessary information). We denote by d_X the Banach-Mazur distance $d(X, \ell_2^{\dim X})$ between a space X and the euclidean space of the same dimension. Throughout this note c denotes a universal positive numerical constant, not necessarily the same in all inequalities.

1. Remarks on Uryshohn's inequality

In the first section we are interested in the behavior of the following functions

$$\varphi(t) = \frac{\mathrm{Vol}(K \backslash tD)}{\mathrm{Vol}K} \quad \text{and} \quad f(t) = \frac{\mathrm{Vol}(K\backslash tD)}{\mathrm{Vol}D} \ ,$$

(where Vol denotes the Lebesgue measure in $\mathbb{R}^n$) and especially how they are concentrated around their respective medians.

It is known that $E(X) = 1$ implies that the median M_X of the function $\|x\|$ on S is around 1 (see [MSch]), which easily implies that $\mathrm{Vol}(D\backslash K) \simeq \mathrm{Vol}(K \cap D)$ (see [M2], section 2.8). Therefore, in this case, for $t < 1$

$$\varphi(t) < ct^n \quad \text{and} \quad f(t) < ct^n$$

for a numerical constant c.

It is known that for some numerical constants c_1 and $c_2 > 0$, if $\varphi(1) < 1/3$ then $\varphi(t) \leq c_1 e^{-c_2 t}$ for $t > 1$ (this follows from C. Borell's lemma [B]; see details in [MSch] Appendix 3).

The Urysohn's inequality [U]

$$\left(\frac{\mathrm{Vol}K}{\mathrm{Vol}D}\right)^{1/n} \leq E^* \tag{1}$$

gives us a control on the volume ratio of K in terms of E^*.

We show below how to estimate $f(t)$ using E^*. As a consequence we will also have another proof of an inequality slightly weaker than (1) with E^* substituted by, say, $2E^*$.

Inequality. *Let $E^* \leq 1$ and let a be such that for any $x \in \mathbb{R}^n$*

$$|x| \leq a\|x\| \ .$$

Then for some numerical sequence $\alpha_n \to 1 (n \to \infty)$

$$\left(\frac{\mathrm{Vol}(K\backslash D)}{\mathrm{Vol}D}\right)^{1/n} \leq \alpha_n \frac{\sin(E^*/a)}{(E^*/a)} E^* \simeq \left(1 - \frac{(E^*)^2}{6a^2}\right) E^* \ .$$

Note that for any norm $\|\cdot\|$, a can be chosen so as to satisfy

$$a \leq \beta_n \sqrt{n} E^*$$

where the numerical sequence $\beta_n \to 1$ when $n \to \infty$ (see[MSch]).

We prove the above inequality assuming $E^* = 1$.

A proof of the following lemma may be found in [M1]. It uses the isoperimetric inequality on S^{n-1}.

Lemma. *Let $A_\varphi = \{x \in S : \|x\| \geq \varphi/E^*\}$.*

Then $\mu(A_\varphi) \ge 1 - c\sqrt{n}\sin^{n-2}\varphi$ for a numerical constant $c > 0$.

Proof of the inequality. Introduce a new euclidean norm $|x|_1 = \theta \cdot |x|$. Then the new unit ball is $D_1 = D/\theta$, $S_1 = \partial D_1$, and $E_1^* = \theta$ (recall that $E^* = 1$). By the Lemma, choosing $\varphi = \theta$, we have

$$\mu\{y \in S_1 : \|y\| \le 1\} < c\sqrt{n}\sin^{n-2}\theta\,. \tag{2}$$

Fix $t > 0$ and consider

$$\frac{\mathrm{Vol}(K \cap (1+t)D_1 \backslash D_1)}{\mathrm{Vol}D_1} = \int_{x \in S_1 \cap K} \left\{\min\left[\frac{1}{\|x\|^n}, (1+t)^n\right] - 1\right\} d\mu(x) \le$$

$$\le [(1+t)^n - 1]\mu\{x \in S_1 \cap K\} \le c\sqrt{n}\{(1+t)^n - 1\}\sin^{n-2}\theta\,,$$

(in the last step we used (2)). Return now to the ball D. Since $\mathrm{Vol}D = \theta^n \mathrm{Vol}D_1$ we have (for $t \ll 1/n$)

$$\frac{\mathrm{Vol}\left(K \cap \frac{1+t}{\theta}D \backslash \frac{1}{\theta}D\right)}{\mathrm{Vol}D} \le cn^{3/2}t\frac{\sin^n\theta}{\theta^n} \cdot \frac{1}{\sin^2\theta}\,. \tag{3}$$

We use (3) $(J+1)$ times for $\theta = \theta_i = 1/(1+t)^i$, $i = 0, 1, \dots, J$, where J is chosen so as to satisfy $(1+t)^J \simeq a$, i.e., $Jt \simeq \ln a$. Then, by (3),

$$\begin{aligned}\frac{\mathrm{Vol}(K \backslash D)}{\mathrm{Vol}D} &\le cn^{3/2}\sum_{i=0}^{J} t\left[(1+t)^i \sin\frac{1}{(1+t)^i}\right]^n \left(\sin\frac{1}{(1+t)^i}\right)^{-2} \le \\ &\le cn^{3/2}(Jt)\left(a\sin\frac{1}{a}\right)^n \left(\sin\frac{1}{a}\right)^{-2}.\end{aligned} \tag{4}$$

□

Application. We call a euclidean structure $|\cdot|$ of a space $(\mathbb{R}^n, \|\cdot\|, |\cdot|)$ *c-suitable* if there exists an orthonormal basis $\{e_i\}_1^n$ such that $\|e_i\| \ge c\sup\|x\|/|x|$ for every $i \le n/2$. (This means that one has many "almost contact" points between the two unit balls after the "right" normalization of the euclidean ball.) For example, the Dvoretzky-Roger's lemma implies that the maximal volume ellipsoid inscribed in the unit ball K induces a c-suitable euclidean structure for, say, $c = \frac{1}{1+\sqrt{2}}$ (see, e.g., [FLM]). Also for every $X = (\mathbb{R}^n, \|\cdot\|)$ there exists a c-suitable euclidean structure, for some absolute constant c, which gives the distance $d(X, \ell_2^n)$ up to a factor of 2 (see [BM2]).

Let $C_q(X^*) = C_q$ be the cotype q-constant of $X^* = (\mathbb{R}^n, \|\cdot\|^*, |\cdot|)$ and let the euclidean structure $|\cdot|$ be α-suitable for X^*. Then the standard computation ([FLM]) shows that

$$\sup_{x \ne 0}|x|/\|x\| = \sup_{x \ne 0}\|x\|^*/|x| \le 2\alpha^{-1}C_q(n^{1/2-1/q})E^*\,.$$

Therefore, by (4), if $E^* = 1$ then

$$\frac{\text{Vol}(K\backslash D)}{\text{Vol}D} \le cn^{5/2}\exp\{-c_1 n^{2/q}/C_q(X^*)^2\}$$

where c_1 depends only on the constant $\alpha > 0$ above.

2. Cotype and Volume Ratio.

Let $\mathcal{E}_K$ be the ellipsoid of maximal volume contained in K, where K is the unit ball of a space $X = (\mathbb{R}^n, \|\cdot\|)$. Define, following [SzT], the volume ratio of a space X

$$vr\ X (= vr\ K) = (\text{Vol}K/\text{Vol}\mathcal{E}_K)^{1/n} .$$

It was proved in [BM1] that $vr\ X \le c \cdot C_2(X)[\log 2C_2(X)]^2$ (here $C_2(X)$ is the Cotype 2-constant of X). This was improved in [MiP] to $vr\ X \le c \cdot C_2(X) \log 2C_2(X)$. Both proofs are quite non-trivial. We demonstrate below a very simple proof of the last inequality.

Let $N(K, \lambda D)$ denote the covering number of K by λD, i.e.,

$$N(K,\lambda D) = \min\{N \in \mathbb{N} \mid \exists x_1, \dots, x_N \in \mathbb{R}^n \text{ and } K \subset \bigcup_1^N (x_i + \lambda D)\} .$$

Let M be the maximal number of disjoint euclidean balls of radius $\lambda/2$ with the centers $\{y_i\}_{i=1}^M$ in K. Then $\{y_i\}$ is a λ-net of K and $N(K, \lambda D) \le M$. Therefore, an obvious volume computation shows that

$$N(K,\lambda D) \le M \le \text{Vol}\left(K + \frac{\lambda}{2}D\right) \Big/ \text{Vol}\left(\frac{\lambda}{2}D\right) .$$

Clearly $E^*_{K+\frac{\lambda}{2}D} = E^*_K + \frac{\lambda}{2}$. Thus, using Urysohn's inequality, we have

$$N(K,\lambda D) \le \left(\frac{E^*_K + \frac{\lambda}{2}}{\lambda/2}\right)^n = e^{n\log\left[1+\frac{2E^*}{\lambda}\right]} .$$

Below we consider the case $\lambda \ge E^*$ and use the inequality in the form

$$N(K,\lambda D) \le e^{cn(E^*/\lambda)} \tag{5}$$

for a universal constant c. (This relation between Urysohn's inequality and the estimation (5) was pointed out to me by A. Pajor [Pa].)

Theorem ([BM1], [MiP]). *Let $C_2(X)$ be the cotype-2-constant of a space $X = (\mathbb{R}^n, \|\cdot\|)$. Then*

$$vr\ X \le c \cdot C_2(X) \log 2C_2(X)$$

where c is a universal constant.

Proof. Consider a space $X = (\mathbb{R}^n, \|\cdot\|, |\cdot|)$ equipped with an euclidean structure $|\cdot|$ such that the unit euclidean ball $D = \{x \in \mathbb{R}^n, |x| \le 1\}$ is the ellipsoid of the maximal volume contained in the unit ball K of X. We use the following two known facts:

i. $E_X^* = E_{K^\circ} \le c_0 \cdot C_2(X) \log 2d_X \le c_0 \cdot C_2(X) \log 2n$ for a universal constant c_0.

(The proof follows from three consequent inequalities: $E_X^* \lesssim T_2(X^*)$ by [DMT], $T_2(X^*) \lesssim C_2(X) \cdot \|\mathrm{Rad}_X\|$ by [MP] and $\|\mathrm{Rad}_X\| \lesssim \log 2d_X$ by [P].)

ii. $N(K, \lambda D) \le \exp(cnE_X^*/\lambda)$ (by (5)).

(We could also use the Sudakov inequality $N(K, \lambda D) \le \exp[cn(E_X^*/\lambda)^2]$.)

By definition of $N = N(K, \lambda D)$ there exist $\{x_i\}_{i=1}^N$ such that

$$K \subset \bigcup_1^N (x_i + \lambda D) \cap K \ .$$

By Brunn's Theorem (see, e.g., [Ch], Lemma 2) $\max_{x \in \mathbb{R}^n} \mathrm{Vol}[(x + \lambda D) \cap K] = \mathrm{Vol}(K \cap \lambda D)$ and therefore

ii′. $\mathrm{Vol} K \le N \cdot \mathrm{Vol}(K \cap \lambda D) \le e^{cnE_X^*/\lambda} \cdot \mathrm{Vol}(K \cap \lambda D)$.

Denote $K_\lambda = K \cap \lambda D$ and let $X_\lambda = (\mathbb{R}^n, \|\cdot\|_\lambda)$ be the normed space with the unit ball $K(X_\lambda) = K_\lambda$. Note that $\|x\|_\lambda = \max(\|x\|, \frac{1}{\lambda}|x|)$ is 2-isomorphic to the ℓ_2-sum of the euclidean norm and $\|\cdot\|$. Therefore,

iii. $C_2(X_\lambda) \le 2C_2(X)$ and $d_{X_\lambda} \le \lambda$.

Note also that the maximal volume ellipsoid of K_λ is again D.

Let $a_1 = \log n$, $a_2 = \log\log n, \dots, a_t = \log^{(t)} n$ - the t-iterated logarithm of n, and let t be the smallest integer such that $\log^{(t)} n < 6c_0$ (here c_0 is the constant in i.).

Take λ_0 such that $K = K \cap \lambda_0 D$ (e.g., $\lambda_0 = \sqrt{n}$) and for $j \ge 1$

$$\lambda_j = a_j E^*_{X_{\lambda_{j-1}}}$$

Then by i. and iii.

$$\lambda_1 \le c_0 C_2(X)(\log 2n)^2$$

and

$$\lambda_j \leq 2c_0 C_2(X) \log(2\lambda_{j-1}) \log^{(j)} n\ .$$

By this one proves, by induction, that

$$\lambda_j \leq 6c_0 \cdot C_2(X)[\log 10C_2(X) + \log^{(j)} n] \log^{(j)} n \tag{6}$$

(for this we need $\log^{(t-1)} n \geq 6c_0$).

We also have, from ii.$'$,

$$\mathrm{Vol}K_{\lambda_{j-1}} \leq e^{cn/a_j} \mathrm{Vol}K_{\lambda_j}\ .$$

Consequently,

$$\mathrm{Vol}K \leq \exp\left[cn\left(\frac{1}{a_1} + \frac{1}{a_2} + \cdots + \frac{1}{a_t}\right)\right] \mathrm{Vol}K_{\lambda_t}\ . \tag{7}$$

Note that, obviously, $\sum_{j=1}^{t} a_j^{-1} < c$ (is uniformly bounded)

$$\mathrm{Vol}K_{\lambda_t} \leq \lambda_t^n \mathrm{Vol}D\ .$$

By (6) $\lambda_t \leq c \cdot C_2(X) \log 10C_2(X)$ for some universal constant c. Thus, we conclude from (7) $vr\ X = (\mathrm{Vol}K/\mathrm{Vol}D)^{1/n} \leq \lambda_t \exp\left[c\left(\frac{1}{a_1} + \frac{1}{a_2} + \cdots + \frac{1}{a_t}\right)\right] \leq c' \cdot C_2(X) \log 10C_2(X)$. □

References

[B] C. Borell. The Brunn-Minkowski inequality in Gauss spaces. Invent. Math. 30 (1975), 207-216.

[BM1] J. Bourgain, V.D. Milman. Sections euclidiennes et volume des corps symetriques convexes dans $\mathbb{R}^n$. C.R. Acad. Sc. Paris, t.300, Ser. 1, N.13 (1985), 435-438.

[BM2] J. Bourgain, V.D. Milman. Distances between normed spaces, their subspaces and quotient spaces. Integral Equations and Operator Theory 9, No. 1 (1986), 31-46.

[Ch] G.D. Chakerian. Inequalities for the difference body of a convex body. Proc. Amer. Math. Soc 18, N. 5 (1967), 879-884.

[DMT] W.J. Davis, V.D. Milman, N. Tomczak-Jaegermann. The diameter of the space of n-dimensional spaces. Israel J. of Math., 39, N. 1-2 (1981), 1-15.

[FLM] T. Figiel, J. Lindenstrauss, V.D. Milman. The dimensions of almost spherical sections of convex bodies. Acta Math 139 (1977), 53-94.

[M1] V.D. Milman. Random subspaces of proportional dimension of finite dimensional normed spaces; approach through the isoperimetric inequality. Missouri Conf. Proceedings, 1984, Springer Lecture Notes, 1166 (1985), 106-115.

[M2] V.D. Milman, Geometrical inequalities and mixed volumes in Local Theory of Banach Spaces. Astérisque, 131 (1985), 373-400.

[MiP] V.D. Milman, G. Pisier. Banach spaces with a weak cotype 2 property. Israel J. Math. 54, N. 2 (1986), 139-158.

[MP] B. Maurey, G. Pisier. Séries de variables aléatories vectorielles indépendantes et propriétés géométriques des espaces de Banach. Studia Math. 58 (1976), 45-90.

[MSch] V.D. Milman, G. Schechtman. Asymptotic theory of finite dimensional normed spaces. Springer Lecture Notes, 1200 (1986).

[Pa] A. Pajor. Sous-espaces ℓ_1^n des espaces de Banach. Thèse de docteur, Université Paris 6, (1984).

[P] G. Pisier. Un théorème sur les opérateurs linéaires entre espaces de Banach qui se factorisent par un espace de Hilbert. Ann. Scient. Ec. Norm. Sup. 13 (1980), 23-43.

[SzT] S.J. Szarek, N. Tomczak-Jaegermann. On nearly Euclidean decompositions for some classes of Banach spaces. Compositio Math. 40 (1980), 367-385.

[U] P.S. Urysohn. Mean width and volume of convex bodies in an n-dimensional space. Mat. Sbornik 31 (1924), 477-486.

ON THE COVERING NUMBERS OF CONVEX BODIES

Hermann König and **Vitali D. Milman**[1]

Mathematisches Seminar
Universität Kiel
23 Kiel, W. Germany

School of Mathematical Sciences
Tel Aviv University
Ramat Aviv, Israel

1. Introduction and results

We study covering numbers of convex bodies, consider the duality problem for them and derive formulas when the convex bodies are in special positions.

If K_1 and K_2 are compact convex, symmetric bodies in $\mathbb{R}^n$, the *covering number*

$$N(K_1, K_2) := \min\{N \in \mathbb{N} \mid \exists y_1, \cdots, y_N \in \mathbb{R}^n \quad K_1 \subset \bigcup_{i=1}^{N} (\{y_i\} + K_2)\}$$

is the minimal cardinality of a covering of K_1 by translates of K_2.

Given a compact, convex, symmetric body $K \subset \mathbb{R}^n$, let $X = (\mathbb{R}^n, \| \, \|)$ denote the normed space whose unit ball is K. Clearly, linear transforms of K define the same Banach space up to isometry. On $\mathbb{R}^n$, we consider an arbitrary but fixed scalar product $(\cdot, \cdot)$ with induced euclidean norm $| \cdot |$. The unit ball of $(\mathbb{R}^n, | \cdot |)$ will be denoted by D. Conversely, any ellipsoid $D \subset \mathbb{R}^n$ defines such a norm. Given D and $| \cdot |$, the dual norm

$$\|x\|^* = \sup\{|(x,y)| \mid \|y\| \leq 1\}$$

has as its unit ball the *polar*

$$K^\circ = \{x \in \mathbb{R}^n \mid |(x,y)| \leq 1 \text{ for all } y \in K\}$$

of the $\| \cdot \|$-unit ball K. The main result of the present note is:

THEOREM 1. *There is a constant $c > 0$ such that for all $n \in \mathbb{N}$, all compact, convex, symmetric bodies K_1, $K_2 \subset \mathbb{R}^n$*

$$c^{-1} N(K_2^\circ, K_1^\circ)^{1/n} \leq N(K_1, K_2)^{1/n} \leq c N(K_2^\circ, K_1^\circ)^{1/n} . \tag{1}$$

[1]Supported in part by the Fund for Basic Research administered by the Israel Academy of Sciences and Humanities.

The essential tools in the proof are inequalities of Satalo and Brunn-Minkowski and recent inverse forms of them. Let K, K_1, K_2 and D be as above. The inequality of *Santalo* [S] states that

$$(\mathrm{vol}_n K)(\mathrm{vol}_n K^\circ) \le (\mathrm{vol}_n D)^2 \ . \tag{2}$$

Here vol_n denotes the n-dimensional volume in $\mathbb{R}^n$. Note that $D = D^\circ$ and thus $(\mathrm{vol}_n D)^2 = (\mathrm{vol}_n D)(\mathrm{vol}_n D^\circ)$. The inequality of *Brunn-Minkowski* [H] reads

$$(\mathrm{vol}_n K_1)^{1/n} + (\mathrm{vol}_n K_2)^{1/n} \le \mathrm{vol}_n(K_1 + K_2)^{1/n} \ . \tag{3}$$

Recently, in [BM] the following converse of (2) was proved.

INVERSE SANTALO INEQUALITY. *There is a constant $c > 1$ such that for all $n \in \mathbb{N}$ and all compact, convex, symmetric bodies $K \subset \mathbb{R}^n$*

$$(\mathrm{vol}_n D)^2 \le c^n (\mathrm{vol}_n K)(\mathrm{vol}_n K^\circ) \ . \tag{4}$$

For a simplified proof see [P_1]. Most importantly, we need the following inverse form of (3) which was recently proved in [M_1]. Another proof is given in [P_2].

INVERSE BRUNN-MINKOWSKI INEQUALITY. *There is a constant $c \ge 1$ such that for all $n \in \mathbb{N}$ and all compact, convex, symmetric bodies $K_1, K_2 \subset \mathbb{R}^n$ there is a linear map $u : \mathbb{R}^n \to \mathbb{R}^n$ with $\det u = 1$ such that for any $\varepsilon, \delta > 0$*

$$\mathrm{vol}_n(uK_1 + \varepsilon K_2)^{1/n} \le c\big((\mathrm{vol}_n K_1)^{1/n} + \varepsilon(\mathrm{vol}_n K_2)^{1/n}\big) \tag{5}$$

and

$$\mathrm{vol}_n\left((uK_1)^\circ + \delta K_2^\circ\right)^{1/n} \le c\big((\mathrm{vol}_n K_1^\circ)^{1/n} + \delta(\mathrm{vol}_n K_2^\circ)^{1/n}\big) \tag{6}$$

We say that uK_1 and K_2 form a *special position* of K_1 and K_2. Note that u does not change volumes. The previous inequalities yield formulas for the covering numbers involving the *volume ratio* of K_1 and K_2,

$$\mathrm{vr}(K_1, K_2) := (\mathrm{vol}_n K_1 / \mathrm{vol}_n K_2)^{1/n} \ .$$

THEOREM 2. *There is a constant $c > 0$ such that for all $n \in \mathbb{N}$ and all compact, convex symmetric bodies $K_1, K_2 \subset \mathbb{R}^n$ there is (a map $u : \mathbb{R}^n \to \mathbb{R}^n$ with $\det u = 1$ and) a special position uK_1 and K_2 of K_1 and K_2 such that for any $\alpha, \beta, \gamma, \delta > 0$ the following is true*

$$\max\left(\mathrm{vr}(K_1, K_2)/\alpha, 1\right) \le N(uK_1, \alpha K_2)^{1/n} \le c \max\left(\mathrm{vr}(K_1, K_2)/\alpha, 1\right) \tag{7}$$

$$(8) \qquad c^{-1} \max\left(\mathrm{vr}(K_1,K_2)/\beta, 1\right) \le N\left(K_2^\circ, \beta(uK_1)^\circ\right)^{1/n} \le c \max\left(\mathrm{vr}(K_1,K_2)/\beta, 1\right)$$

$$(9) \qquad \max\left(\mathrm{vr}(K_2,K_1)/\gamma, 1\right) \le N\left(K_2, \gamma u K_1\right)^{1/n} \le c \max\left(\mathrm{vr}(K_2,K_1)/\gamma, 1\right)$$

$$(10) \qquad c^{-1} \max\left(\mathrm{vr}(K_2,K_1)/\delta, 1\right) \le N\left((uK_1)^\circ, \delta K_2^\circ\right)^{1/n} \le c \max\left(\mathrm{vr}(K_2,K_1)/\delta, 1\right) .$$

Theorem 1 is proved using these formulas. Note that if K_1 and K_2 are already in a special position, i.e., $u = id$, (1) follows directly from (7) and (8). In particular, if $\mathrm{vol}_n K_1 = \mathrm{vol}_n K_2$, we get immediately

COROLLARY. *There is $c > 0$ such that for all $n \in \mathbb{N}$ and all $K_1, K_2 \subset \mathbb{R}^n$ as above with $\mathrm{vol}_n K_1 = \mathrm{vol}_n K_2$ there is a special position uK_1 and K_2 such that*

$$(11) \qquad 1 \le [N(uK_1, K_2) \cdot N((uK_1)^\circ, K_2^\circ)]^{1/n} \le c .$$

We now give an operator-theoretic interpretation of Theorem 1 in terms of the *entropy numbers* $e_n(v)$ of an operator $v : X \to Y$ between (real) Banach spaces X and Y,

$$e_j(v) := \inf\left\{\varepsilon > 0 \mid \exists y_1, \ldots, y_{2^{j-1}} \in Y \quad v(K_X) \subset \bigcup_{i=1}^{2^{j-1}} (\{y_i\} + \varepsilon K_Y)\right\} .$$

Here K_X and K_Y denote the closed unit balls of X and Y, respectively. These numbers are essentially an inverse of the covering numbers since $\varepsilon \underset{(-)}{>} e_j(v)$ iff $N(vK_X, \varepsilon K_Y) \le 2^{j-1}$. Note that:

$$v \text{ is compact} \iff e_n(v) \to 0 \iff e_n(v^*) \to 0 .$$

Pietsch posed the question whether for any operator $v : X \to Y$ these numbers are essentially selfdual, i.e., whether

$$0 < \inf_{j \in N} \frac{e_j(v)}{e_j(v^*)} \le \sup_{j \in \mathbb{N}} \frac{e_j(v)}{e_j(v^*)} < \infty .$$

An (almost) positive answer was given in [GKS] in the case that X and Y^* are of type 2. This was extended to the case where X^* and Y are of cotype 2 in [PT_2]. Further partial positive answers are found in [PT_1]. At the time of writing, the general case is still unsolved. However, Theorem 1 yields the following positive information on the problem.

THEOREM 3. *For any $\lambda > 0$ there is $d = d(\lambda) > 1$ such that for any finite rank operator $v : X \to Y$ between (real) Banach spaces and all $j > \lambda \cdot \mathrm{rank}\, v$ one has*

$$(12) \qquad e_{[dj]}(v) \le 2e_j(v^*) \quad , \quad e_{[dj]}(v^*) \le 2e_j(v) .$$

For $\lambda \to \infty$ one has $d(\lambda) \to 1$.

2. Proofs of the previous theorems

Proof of Theorem 2

Take any $u : \mathbb{R}^n \to \mathbb{R}^n$ with $\det u = 1$. By definition of $N = N(uK_1, \alpha K_2)$ there are $y_1, \dots, y_N \in \mathbb{R}^n$ with

$$uK_1 \subset \bigcup_{i=1}^{N} (\{y_i\} + \alpha K_2) .$$

Taking volumes, we find the left inequality in (7):

$$\mathrm{vol}_n K_1 = \mathrm{vol}_n(uK_1) \le N\alpha^n \,\mathrm{vol}_n K_2 , \quad \mathrm{vr}(K_1, K_2)/\alpha \le N^{1/n} .$$

A similar argument yields that

$$\mathrm{vr}\big(K_2^\circ, (uK_1)^\circ\big)/\beta \le N\big(K_2^\circ, \beta(uK_1)^\circ\big) .$$

By Santalo's inequality (2) and its inverse (4),

$$(\mathrm{vol}_n K_2^\circ)(\mathrm{vol}_n K_2) \ge (\mathrm{vol}_n D)^2$$
$$\big(\mathrm{vol}_n (uK_1)^\circ\big)(\mathrm{vol}_n uK_1) \le c^n (\mathrm{vol}_n D)^n .$$

Thus $c^{-1}\mathrm{vr}(K_1, K_2) = c^{-1}\mathrm{vr}(uK_1, K_2) \le \mathrm{vr}\big(K_2^\circ, (uK_1)^\circ\big)$, which proves the left side of (8). For the left inequalities in (9) and (10) similar proofs apply.

To prove the right-hand inequalities of (7)-(10), choose a special position of K_1 and K_2, i.e., a map $u : \mathbb{R}^n \to \mathbb{R}^n$ with $\det u = 1$ such that (5) and (6) hold with constant $c \ge 1$. Let $\mathcal{N} = \big\{\{y_i\} + \alpha/2\ K_2 \mid y_i \in uK_1 \ , \ \forall i \ne j \ \ (\{y_i\} + \alpha/2\ K_2) \cap (\{y_j\} + \alpha/2\ K_2) = \emptyset\big\}$ be a collection of disjoint translates of $\alpha/2\ K_2$ with centers in uK_1 having maximal cardinality $N = |\mathcal{N}|$ among all such collections. Clearly

$$u(K_1) \subset \bigcup_{i=1}^{N} (\{y_i\} + \alpha K_2)$$

because of the maximality of $\mathcal{N}$ and thus $N(uK_,\alpha K_2) \le N$. Moreover,

$$\bigcup_{i=1}^{N} (\{y_i\} + \alpha/2\ K_2) \subset uK_1 + \alpha/2\ K_2 .$$

Since these balls are disjoint, we find taking volumes

$$N \cdot \mathrm{vol}_n(\alpha/2\,K_2) \leq \mathrm{vol}_n(uK_1 + \alpha/2\,K_2)\ .$$

By (5)

$$\begin{aligned} N(uK_1, \alpha K_2)^{1/n} \leq N^{1/n} &\leq c\big(2/\alpha\,\mathrm{vr}(K_1, K_2) + 1\big) \\ &\leq 3c \cdot \max\big(\mathrm{vr}(K_1, K_2)/\alpha, 1\big) \end{aligned}$$

which proves (7). Concerning (10), a similar argument yields

$$N\big((uK_1)^\circ, \delta K_2^\circ\big)^{1/n} \leq [\mathrm{vol}_n\,\big((uK_1)^\circ + \delta/2\,K_2^\circ\big) / \mathrm{vol}_n(\delta/2\,K_2^\circ)]^{1/n}$$

which by (6) is bounded by

$$\begin{aligned} &\leq c\big(2/\delta\,\mathrm{vr}((uK_1)^\circ, K_2^\circ) + 1\big) \\ &\leq 3c\max\big(\mathrm{vr}((uK_1)^\circ, K_2^\circ)/\delta, 1\big) \end{aligned}$$

Another application of Santalo's inequality and its inverse yields

$$\mathrm{vr}\big((uK_1)^\circ, K_2^\circ\big) \leq d\,\mathrm{vr}(K_2, uK_1) = d\,\mathrm{vr}(K_2, K_1)$$

with some absolute constant $d \geq 1$. This proves (10).

Choosing a maximal collection of disjoint translates of $\gamma/2\,uK_1$ with centers in K_2, we find using (5) again

$$\begin{aligned} N(K_2, \gamma uK_1)^{1/n} &\leq \big(\mathrm{vol}_n(K_2 + \frac{\gamma}{2}uK_1) / \mathrm{vol}_n(\frac{\gamma}{2}uK_1)\big)^{1/n} \\ &= \big(\mathrm{vol}_n(uK_1 + \frac{2}{\gamma}K_2) / \mathrm{vol}_n(uK_1)\big)^{1/n} \\ &\leq c\big(1 + \frac{2}{\gamma}\mathrm{vr}(K_2, K_1)\big) \\ &\leq 3c\max\big(\mathrm{vr}(K_2, K_1)/\gamma, 1\big)\ . \end{aligned}$$

This proves (9). The right side of (8) uses additionally again Santalo's inequality and its inverse. □

Formula (11) is a direct application of (7) and (10) with $\mathrm{vol}_n\,K_1 = \mathrm{vol}_n\,K_2$ and $\alpha = \delta = 1$.

Proof of Theorem 1

We prove the right inequality of (1) with constant c^2, where c is the constant in formula (11) (of the corollary to Theorem 2). The obvious subadditivity property of the covering numbers

$$N(K_1, K_3) \leq N(K_1, K_2) N(K_2, K_3)$$

will be used in the following. Moreover $N(K_1, K_2) = N(vK_1, vK_2)$ for $v : \mathbb{R}^n \to \mathbb{R}^n$ invertible. Let K_1, K_2 be as in Theorem 1, and define α and β by

$$\mathrm{vol}_n\, K_1 = \mathrm{vol}_n\, \alpha D \ , \ \mathrm{vol}_n\, K_2 = \mathrm{vol}_n\, \beta D \ .$$

According to the corollary to Theorem 2, there is a special position $u_1 K_1$ and αD of K_1 and αD such that

$$1 \le N(u_1 K_1, \alpha D) N\big((u_1 K_1)^\circ, (\alpha D)^\circ\big) \le c^n \tag{13}$$

and a special position $u_2(\beta D)$ and K_2 of βD and K_2 such that

$$1 \le N(u_2 \beta D, K_2) N\big((u_2 \beta D)^\circ, \beta K_2^\circ\big) \le c^n \ . \tag{14}$$

Here $u_1, u_2 : \mathbb{R}^n \to \mathbb{R}^n$ are maps with $\det u_1 = \det u_2 = 1$. Note that $D^\circ = D$. We have $(u_1 K_1)^\circ = u_1^{*-1} K_1^\circ$ and $(u_2 \beta D)^\circ = u_2^{*-1}(\beta^{-1} D)$ and thus $N(u_2^{*-1}\beta^{-1}D, K_2^\circ) = N\big((u_2\beta D)^\circ, K_2^\circ\big)$ Further,

$$N(K_1, u_1^{-1}\alpha D) = N(u_1 K_1, \alpha D)$$

and

$$N(K_1^\circ, u_1^* \alpha D) = N(u_1^{*-1} K_1^\circ, \alpha D) = N\big((u_1 K_1)^\circ, \alpha D\big) \ .$$

Using this, subadditivity yields

$$\begin{aligned} N(K_1, K_2) &\le N(K_1, u_1^{-1}\alpha D) N(u_1^{-1}\alpha D, u_2 \beta D) N(u_2 \beta D, K_2) \\ &= N(u_1 K_1, \alpha D) N(u_2 \beta D, K_2) \cdot N(u_1^{-1}\alpha D, u_2 \beta D) \ , \\ N(u_2^{*-1}\beta^{-1}D, u_1^*\alpha^{-1}D) &\le N(u_2^{*-1}\beta^{-1}D, K_2^\circ) N(K_2^\circ, K_1^\circ) N(K_1^\circ, u_1^*\alpha^{-1}D) \\ &= N\big((u_2\beta D)^\circ, K_2^\circ\big) N\big((u_1 K_1)^\circ, (\alpha D)^\circ\big) \cdot N(K_2^\circ, K_1^\circ) \ . \end{aligned}$$

We claim that

$$N(u_1^{-1}\alpha D, u_2 \beta D) = N(u_2^{*-1}\beta^{-1}D, u_1^*\alpha^{-1}D) \ . \tag{15}$$

Using this, the previous estimates, (13) and (14) imply

$$\begin{aligned} N(K_1, K_2) &\le N(u_1 K_1, \alpha D) N\big((u_1 K_1)^\circ, (\alpha D)^\circ\big) \\ &\qquad \cdot N(u_2 \beta D, K_2) N\big((u_2 \beta D)^\circ, K_2^\circ\big) \cdot N(K_2^\circ, K_1^\circ) \\ &\le c^{2n} N(K_2^\circ, K_1^\circ) \ . \end{aligned}$$

This will prove (1). It remains to check (15). Let $v := \frac{\alpha}{\beta}(u_1 u_2)^{-1}$. Then (15) is equivalent to

$$N(vD, D) = N(v^* D, D) \ . \tag{16}$$

By polar decomposition in the Hilbert space $H = (\mathbb{R}^n, |\cdot|)$ with unit ball D there is an isometry $U : H \to H$ and a positive map $p = p^* : H \to H$ (square-root of $v^* v$) such that $v = Up$. Thus $UD = U^{-1}D = D$ and $v^* = pU^*$. Hence

$$\begin{aligned} N(vD, D) &= N(pD, U^{-1}D) = N(pD, D) \\ &= N(pU^* D, D) = N(v^* D, D) \ , \end{aligned}$$

which proves (16). □

Proof of Theorem 3

Let $v : X \to Y$ have finite rank n. Consider the standard factorizations of v and v^*, factoring out the kernels,

$$\begin{aligned} &v \colon X \xrightarrow{\pi} X/\mathrm{Ker}\, v \xrightarrow{\overline{v}} \mathrm{Im}\, v \overset{i}{\hookrightarrow} Y \\ &v^* \colon y^* \xrightarrow{\tilde{\pi}} Y^*/\mathrm{Ker}\, v^* = (\mathrm{Im}\, v)^* \xrightarrow{\overline{v}^*} (X/\mathrm{Ker}\, v)^* = \mathrm{Im}\, v^* \overset{\tilde{i}}{\hookrightarrow} X^* \ . \end{aligned}$$

By elementary properties of the entropy numbers, one has

$$e_j(v) \le e_j(\overline{v}) \le 2e_j(v) \quad , \quad e_j(v^*) \le e_j(\overline{v}^*) \le 2e_j(v^*) \ .$$

This shows that to prove (12), it suffices to show

$$e_{[d(\lambda)j]}(v) \le e_j(v^*) \quad , \quad e_{[d(\lambda)j]}(v^*) \le e_j(v)$$

for all $v : X \to Y$ between n-dimensional spaces and $j > \lambda n$. We prove this with $d = d(\lambda) = 1 + a/\lambda$, where $a := \log_2 c$ and $c > 1$ is the constant in Theorem 1. Clearly then, $\lim_{\lambda \to \infty} d(\lambda) = 1$. Without loss of generality we may assume that v is invertible (density argument). Let K_X and K_Y denote the unit balls in X and Y, respectively. We put $K_1 := v(K_X)$ and $K_2 := K_Y$. Then $K_1^\circ = v^{*-1}(K_{X^*})$, $K_2^\circ = K_{Y^*}$. Clearly

$$N(K_2^\circ, \varepsilon K_1^\circ) = N(K_{Y^*}, \varepsilon v^{*-1} K_{X^*}) = N(v^* K_{Y^*}, \varepsilon K_{X^*}) \ .$$

Thus by Theorem 1

$$c^{-n} N(v^* K_{Y^*}, \varepsilon K_{X^*}) \le N(vK_X, \varepsilon K_Y) \le c^n N(v^* K_{Y^*}, \varepsilon K_{X^*})$$

which implies with $a := \log_2 c$ that

(17) $$\varepsilon_{j+[an]}(v^*) \le \varepsilon_j(v) \quad , \quad \varepsilon_{j+[an]}(v) \le \varepsilon_j(v^*) \ .$$

For $j > \lambda n$, $j + an < (1 + a/\lambda) \ \ j = d(\lambda)j$. Thus

$$e_{[dj]}(v^*) \le \varepsilon_j(v) \quad , \quad \varepsilon_{[dj]}(v) \le \varepsilon_j(v^*) \ .$$

□

Remarks about the special position. We say that compact, convex, symmetric bodies $K_1, K_2 \subset I\!R^n$ are already in a special position if the map u in (5)-(6) [and thus also (7)-(10)] can be taken to be the identity map $u = id$.

i) Let X be of cotype 2, K_X be its unit ball the D the maximal volume ellipsoid inside K_X. Then K_X and D are already in a special position. More exactly, this means that inequalities (5) and (6) hold with $u = id$ and constant c depending only on the cotype 2-constant $C_2(X)$ of X only. This follows from the proof in [BM] that cotype 2-spaces have finite volume ratio. Another proof of this fact which also implies the above statement is given in [MP].

ii) For the same reason, if X and Y have cotype 2 and the maximal volume ellipsoids of K_X and K_Y are proportional, then K_X and K_Y are already in a special position, i.e., (5)-(6) holds with $u = id$ and $c = c(C_2(X), C_2(Y))$.

iii) Consider the dual case when X^* and Y^* are of cotype 2. Then the minimal volume ellipsoid circumscribed to K_X is the maximal volume ellipsoid inside $(K_X)^\circ = K_{X^*}$. Thus, if any such ellipsoids for K_X and K_Y are proportional, these bodies K_X and K_Y are already in a special position.

iv) Finally, consider the case that X and Y^* have cotype 2. If the maximal volume ellipsoid for K_X is proportional to the minimal volume ellipsoid for K_Y, then K_X and K_Y are already in a special position.

In all these cases, the estimates (7)-(10) for covering numbers hold for $u = id$ and c depending only on the cotype 2-constants of the spaces involved (with unit balls $K_1, K_2, K_1^\circ, K_2^\circ$, respectively).

3. Covering numbers in subspaces and quotients

If K is a compact, convex, symmetric body in $I\!R^n$ and $X = (I\!R^n, \|\cdot\|)$ is the normed space with unit ball K, by [M$_2$] there is a quotient of a subspace qsX of X with dimension almost n

($\dim qsX \geq n - c\frac{n}{(\ln d_K)^2}$) such that the unit ball K_1 of qsX is much closer to an euclidean ball than K, say

$$d_{K_1} \leq c(\log 2d_K)^2 . \tag{18}$$

We now show that the covering numbers (or the entropy numbers) have a "weak dependence" on the dimension, i.e., if there is only a small change in the dimension, the covering numbers will not change too much by a "quotient of a subspace" operation qs as indicated above. That a similar fact holds for the volume is used strongly in [BM].

The following result together with (18), applied finitely many times, reduces the approximate calculation of covering numbers of convex bodies to covering numbers of ellipsoids which, of course, is easier. Actually, Theorem 4 provides an alternative possibility for proving Theorems 1-3, since the self-dual property of the covering numbers holds for ellipsoids (see(16)).

THEOREM 4. *Let K_1 and K_2 be compact, convex, symmetric bodies in $\mathbb{R}^n$ such that $b^{-1}K_2 \subset K_1 \subset aK_2$ for some $a, b > 0$. Let $X = (\mathbb{R}^n, \|\cdot\|)$ be the normed space with the unit ball K_2, $E \subset X$ be a subspace of codimension j and $F \subset E$ be a subspace of dimension j and $\pi : E \to E/F$ be the quotient map. Denote $qsK_1 := \pi(K_1 \cap E)$, $qs(K_2) := \pi(K_2 \cap E)$. Then for any $0 < \varepsilon \leq b^{-1}$*

$$d_1^{-j}\varepsilon^{-2j}N(qsK_1, c\varepsilon\, qsK_2) \leq N(K_1, \varepsilon K_2) \leq d_2^{-j}\varepsilon^{-2j}N(qsK_1, c^{-1}\varepsilon\, qsK_2) .$$

Here $c > 1$ is a numerical constant and

$$d_1 := c{d_{K_2}}^3a^2 \quad , \quad d_2 := c{d_{K_2}}^3ab^3$$

where d_{K_2} denotes the Banach-Mazur distance of X to ℓ_2^n.

The term "weak dependence of the covering numbers on the dimension" can be made more precise by noting that if $ab \lesssim d_{K_2}$ and $2j =$ dimensional change $\sim n/(\ln d_{K_2})^\alpha$ for suitable $\alpha > 0$, then

$$\left(N(K_1, K_2)/N(qsK_1, c^{-1}\varepsilon\, qsK_2)\right)^{1/n} \leq c\varepsilon^{-1/(\ln d_{K_2})^\alpha} .$$

Theorem 4 follows directly from Propositions 1-3 below. We start by comparing affine sections of convex bodies with linear sections.

LEMMA. *Let K be a compact, convex, symmetric body in $\mathbb{R}^n$ and $E \subset \mathbb{R}^n$ be a subspace. Then for every $y_0 \in \mathbb{R}^n$ there is $y_1 \in \{y_0\} + E$ such that*

$$(\{y_0\} + E) \cap K \subset \{y_1\} + 2(K \cap E) \tag{19}$$

If for some $0 < \varepsilon < 1$ the K-distance of y_0 to E is at most ε, i.e., if $(\{y_0\} + \varepsilon K) \cap E \neq \emptyset$, this can be improved to

$$(\{y_0\} + E) \cap K \subset \{y_1\} + (1+\varepsilon)(K \cap E).$$

Proof Let $0 < \varepsilon \leq 1$, $y_0 \in \mathbb{R}^n$ and $F = \operatorname{span}(y_0, E)$. We may assume that $y_0 \notin E$ and $(\{y_0\} + E) \cap K \neq \emptyset$. Choose $y^* \in F^*$ such that $E = \operatorname{Ker} y^*$ and $y^*(y_o) \geq 0$. Since $K_F := K \cap F$ is compact, there is $y_2 \in K_F$ such that $y^*(y_2) = \inf_{y \in K_F} y^*(y)$. Thus $y^*(y_2) < 0$. Define $y_1 := y^*(y_0)/y^*(y_2) \cdot y_2$. Clearly $y_1 \in K$ and $y_1 - y_0 \in \operatorname{Ker} y^* = E$. Now take any $z \in (\{y_0\} + E) \cap K$. Since K is convex,

$$y_3 := (|y^*(y_2)| z + y^*(z) y_2)/(|y^*(y_2)| + y^*(z)) \in K\ .$$

Moreover $y_3 \in \operatorname{Ker} y^* = E$. Since $(\{y_0\} + \varepsilon K) \cap E \neq \emptyset$, there is $y_4 \in K$ with $y^*(y_0) + \varepsilon y^*(y_4) = 0$. By definition of y_2, $y^*(y_4) \geq y^*(y_2)$ and thus

$$y^*(z) = y^*(y_0) = -\varepsilon y^*(y_4) \leq -\varepsilon y^*(y_2) = \varepsilon |y^*(y_2)|\ .$$

We find

$$z - y_1 = [(|y^*(y_2)| + y^*(z))/|y^*(y_2)|] y_3 \in (1+\varepsilon)(K \cap E)\ .$$

□

COROLLARY. *(Maximum principle for covering numbers).*

Let K_1 and K_2 be compact, convex, symmetric bodies in $\mathbb{R}^n$ and E be a subspace of $\mathbb{R}^n$. Then for every $y_0 \in \mathbb{R}^n$ the following holds

$$N(E \cap K_1, K_2) \geq N\big((\{y_0\} + E) \cap K_1, 2K_2\big)$$

Proof Immediate from (19). □

PROPOSITION 1. *Let K_1 and K_2 be compact, convex symmetric bodies in $\mathbb{R}^n$ such that $b^{-1} K_2 \subset K_1 \subset a K_2$ for some $a, b > 0$. Let E be a subspace of codimension j. Then for any $\varepsilon > 0$ and $0 < \delta \leq b^{-1}$*

$$N(K_1, (4\varepsilon + \delta) K_2) \leq (3 d_{K_2}\, a/\delta)^j N(K_1 \cap E, \varepsilon(K_2 \cap E))$$

Let $X_2 = (I\!R^n, \|\cdot\|)$ be the Banach space with unit ball K_2 . Then d_{K_2} denotes the Banach-Mazur distance of X_2 to ℓ_2^n.

Proof. Let $|\cdot|$ be an euclidean norm on $I\!R^n$ such that $|x| \leq \|x\| \leq d_{K_2}|x|$ for all $x \in I\!R^n$ and let $E^\perp$ be the $|\cdot|$-orthogonal complement of E in $I\!R^n$. By (19), applied to $K_1 + \delta K_2$, for any $y_0 \in E^\perp$ there is $y_1 \in \{y_0\} + E$ such that

$$(\{y_0\} + E) \cap (K_1 + \delta K_2) \subset \{y_1\} + 2(K_1 + \delta K_2) \cap E$$

$$\subset \{y_1\} + 2(1 + \delta b)K_1 \cap E \subset \{y_1\} + 4(K_1 \cap E) .$$

Let $K := K_2 \cap E$ and $K^\perp := K_2 \cap E^\perp$. Then there is a δ-net $\mathcal{N}_\delta$ in $E^\perp$ for $d_{K_2} \cdot a \cdot K^\perp$ with respect to the norm $\|\cdot\|$ which has cardinality $|\mathcal{N}_\delta| \leq (3d_{K_2}a/\delta)^j$ since $\dim E^\perp = j$. By definition of $N := N(K_1 \cap E, \varepsilon(K_2 \cap E))$ there is $\mathcal{M}_\varepsilon = \{z^1, \ldots, z^N\} \subset E$ such that

$$K_1 \cap E \subset \bigcup_{k=1}^{N} (\{z^k\} + \varepsilon(K_2 \cap E))$$

For each $y_0{}^i \in \mathcal{N}_\delta$, $i = 1, \ldots, |\mathcal{N}_\delta|$ choose $y_1{}^i \in \{y_0{}^i\} + E$ such that

$$(\{y_0{}^i\} + E) \cap (K_1 + \delta K_2) \subset \{y_1{}^i\} + 4(K_1 \cap E) .$$

Let $\mathcal{L} := \{y_1{}^i + 4z^k \mid i = 1, \ldots, |\mathcal{N}_\delta|,\ k = 1, \ldots, N\}$. We claim that

$$K_1 \subset \mathcal{L} + (4\varepsilon + \delta)K_2 . \tag{20}$$

Then obviously

$$N\big(K_1, (4\varepsilon + \delta)K_2\big) \leq |\mathcal{L}| \leq (3d_{K_2}\ a/\delta)^j \cdot N .$$

To prove (20), let $x \in K_1$, $x = z + y$, $z \in E$, $y \in E^\perp$. Clearly

$$\|y\| \leq d_{K_2}|y| \leq d_{K_2}|x| \leq d_{K_2}\|x\| \leq d_{K_2}a$$

since $K_1 \subset aK_2$. Hence $y \in d_{K_2}aK^\perp$ and there is $y_0{}^i \in \mathcal{N}_\delta$ with

$$\|y - y_o{}^i\| = \|x - (z + y_o^i)\| \leq \delta .$$

Since $z + y_0{}^i \in (\{y_0{}^i\} + E) \cap (K_1 + \delta K_2) \subset \{y_1{}^i\} + 4(K_1 \cap E)$, we find $z + y_0{}^i - y_1{}^i \in 4(K_1 \cap E)$ and there is $z^k \in \mathcal{M}_\varepsilon$ with

$$\|z + y_0{}^i - y_1{}^i - 4z^k\| \leq 4\varepsilon ,$$

$$\|x - (y_1{}^i + 4z^k)\| \leq \|x - (z + y_o{}^i)\| + \|(z + y_0{}^i) - y_1{}^i - 4z^k\| \leq 4\varepsilon + \delta ,$$

i.e., (20) holds. □

A dual version of Proposition 1 is

PROPOSITION 2. *Let K_1, K_2, a, b be as before, $X_2 = (\mathbb{R}^n, \|\cdot\|)$ be the Banach space with unit ball K_2, $E \subset \mathbb{R}^n$ be a subspace and $\pi : X_2 \to X_2/E$ be the quotient map. Then for any $\varepsilon > 0$ and $o < \delta \leq b^{-1}$*

$$N(K_1, (16\varepsilon + \delta)K_2) \leq (9{d_{K_2}}^2 a/\delta)^j N(\pi K_1, \varepsilon\pi K_2) .$$

Proof. Let $|\cdot|$ and $E^\perp$ be as before and identify $E^\perp$ algebraically with X/E via π. By definition, there is $\mathcal{M}_\varepsilon \subset X/E$ with $|\mathcal{M}_\varepsilon| = N(\pi K_1, \varepsilon\pi K_2)$ such that

$$\pi K_1 \subset \mathcal{M}_\varepsilon + \varepsilon\pi K_2 .$$

Let $\eta = {d_{K_2}}^{-1}$, K be the $|\cdot|$-unit ball, and choose an $|\cdot|$-η-net $\mathcal{N}_\eta$ of cardinality $|\mathcal{N}_\eta| \leq (3d_{K_2})^j$ for $K \cap E$. We will cover πK_2 by a "small" number of balls $4\pi(K_2 \cap E^\perp)$. Thus let $\overline{y} \in \pi \overset{\circ}{K}_2$, take $y \in E^\perp$ with $\pi y = \overline{y}$ and choose $e \in E$ with $\|y - e\| < 1$. Since $|e| \leq |y - e| \leq \|y - e\| < 1$, $-e \in K \cap E$. Hence there is ${y_o}^i \in \mathcal{N}_\eta$ with $|{y_o}^i + e| < \eta$. Then

$$\|y + {y_o}^i\| \leq \|y - e\| + \|{y_o}^i + e\| < 1 + d_{K_2}|{y_o}^i + e| < 2 ,$$

${y_o}^i + y \in (\{{y_o}^i\} + E^\perp) \cap 2K_2$. By (19) of the lemma, ${y_o}^i + y \in \{{y_1}^i\} + 4(K_2 \cap E^\perp)$ for some ${y_1}^i \in \{{y_o}^i\} + E^\perp$ determed by ${y_o}^i$. If $P : \mathbb{R}^n \to E^\perp$ denotes the $|\cdot|$-orthogonal projection,

$$y = P({y_o}^i + y) \in \{P{y_1}^i\} + 4(K_2 \cap E^\perp) .$$

Thus

$$\pi K_2 \subset \bigcup_{{y_o}^i \in \mathcal{N}_\eta} \{P{y_1}^i\} + 4\pi(K_2 \cap E^\perp)$$

$$\pi K_1 \subset \bigcup_{{y_o}^i \in \mathcal{N}_\eta, z^k \in \mathcal{M}_\varepsilon} (\{P y_1^i + z^k\} + 4\varepsilon\pi(K_2 \cap E^\perp))$$

which - using Proposition 1 - yields

$$N(K_1,(16\varepsilon+\delta)K_2) \le (3d_{K_2}\, a/\delta)^j N(K_1\cap E^\perp, 4\varepsilon(K_2\cap E^\perp))$$
$$\le (9{d_{K_2}}^2\, a/\delta)^j N(\pi K_1, \varepsilon\pi K_2)\ .$$

□

Propositions 1 and 2 give upper estimates for covering numbers by covering numbers in subspaces or quotient spaces. Converse lower estimates can be proved by volume arguments; a covering of convex bodies in $\mathbb{R}^n$ requires "much" more balls than corresponding covering in subspaces. One can prove a partial converse of Proposition 1 and 2.

PROPOSITION 3. *Let K_1 and K_2 be compact, convex, symmetric bodies in $\mathbb{R}^n$ and $b^{-1}K_2 \subset K_1 \subset aK_2$ for suitable $a,b>0$. Let $X_2=(\mathbb{R}^n,\|\cdot\|)$ be the Banach space with unit ball K_2, $E\subset X_2$ be a subspace of codimension j, F be a subspace of dimension j and $\pi: X\to X/F$ be the quotient map. Then for any $\varepsilon>0$*

$$N\left(K_1,\frac{\varepsilon}{2}K_2\right) \ge \left(\frac{c}{d_{K_2}b\varepsilon}\right)^j N(K_1\cap E, \varepsilon(K_2\cap E))$$
$$N\left(K_1,\frac{\varepsilon}{8}K_2\right) \ge \left(\frac{c}{{d_{K_2}}^2ab^2\varepsilon}\right)^j N(\pi K_1, \varepsilon\pi K_2)$$

where $c\ge 1$ is an absolute constant.

References

[BM] J. Bourgain, V.D. Milman; Sections euclidiennes et volume des corps symmetriques convexes dans $\mathbb{R}^n$, C.R. Acad. Sc. Paris, t.300 Serie 1, N. 13 (1985) 435-438. (See also: On Mahler's conjecture on the volume of a convex symmetric body and its polar, (Preprint I.H.E.S., 1985).

[GKS] Y. Gordon, H. König, S. Schütt; Geometric and probabilistic estimates for entropy and approximation numbers of operators, to appear in J. Appr. Th.

[H] H. Hadwiger; Vorlesungen Über Inhalt: Oberfläche und Isoperimetrie, Springer-Verlag, 1957.

[M$_1$] V.D. Milman; Inegalité de Brunn-Minkowski inverse et applications à la théorie locale des espaces normés, CRAS 302 (1986), 25-28.

[M$_2$] V.D. Milman; Almost Eulidean quotient spaces of subspaces of finite dimensional normed spaces, Proc. AMS 94 (1985), 445-449.

[P_1] G. Pisier; Private communication.

[P_2] G. Pisier; A simpler proof of several results of V. Milman.

[PT_1] A. Pajor, N. Tomczak-Jaegermann; Remarques sur les nombres d'entropie d'un opérateur et de son transposé, CRAS

[PT_2] A. Pajor, N. Tomczak-Jaegermann; in preparation

[S] L.A. Santalo; Un invariante afin pasa los cuerpos convexos del espacio de n dimensiones, Portugal Math. 8 (1949), 155-161.

ON A THEOREM OF J. BOURGAIN ON FINITE DIMENSIONAL DECOMPOSITIONS AND THE RADON-NIKODYM PROPERTY

W. Schachermayer
Department of Mathematics
University of Linz
Austria

Abstract: We give an exposition of the "hard case" of Bourgain's theorem, that a Banach space X has RNP iff each subspace with a finite dimensional decomposition has RNP. We reproduce essentially Bourgain's arguments, by explaining the ideas underlying the proof and giving slightly altered arguments for some of the technical details.

1. Introduction: A real Banach space X has the Radon-Nikodym property (RNP) (c.f. [D-U]) if for every closed, convex, bounded set $C \subseteq X$ and $\varepsilon > 0$ there is a slice S of C of diameter less then ε. A slice of C is a set of the form

$$S = S(x^*, \beta) = \{x \in C : \langle x^*, x\rangle > M_{x^*} - \beta\} ,$$

where $\beta > 0$, x^* is in the unit sphere of X^* and

$$M_{x^*} = \sup\{\langle x^*, z\rangle : z \in C\} .$$

In the last years some relatives of the notion of (RNP) became important for a deeper understanding of RNP:

Definition 1.1: a) X has the *Point of Continuity property* (PCP) if for every weakly closed, bounded $F \subseteq X$ there is a point of continuity of the identity map $(F, \text{weak}) \mapsto (F, \|\cdot\|)$.

b) X has the *convex Point-of-Continuity property* (CPCP) if for every convex, closed, bounded $C \subseteq X$ there is a point of continuity of the identity map $(C, \text{weak}) \mapsto (C, \|\cdot\|)$.

c) X is *strongly regular* (SR) if for every convex, closed, bounded $C \subseteq X$ and for every $\varepsilon > 0$ there are slices $S_1, \ldots, S_n$ of C s.t.

$$\text{diameter}\left(n^{-1}(S_1 + \ldots + S_n)\right) < \varepsilon .$$

The notion of (CPCP) - obviously a weakening of the concept of (RNP) - was introduced by J. Bourgain (under the name property (*)) in [B1] in order to solve the problem of f.d.d.'s we are dealing with here. Let us formulate precisely the theorem that we are going to prove.

Theorem 1.2 [B1]: *Let X be a Banach space failing (RNP) and $\lambda > 1$. There is a subspace E of X and a uniformly bounded, E-valued, simple-valued martingale $(\xi_n)_{n=0}^{\infty}$ on $[0,1]$ and a sequence of finite rank projections $S_n : E \mapsto E$ such that*

1. $x = \lim S_n(x)$ *for* $x \in E$
2. $\|S_n\| \leq \lambda$
3. $S_m S_n = S_n S_m = S_m$ *for* $m \leq n$
4. $S_n \xi_{n+1} = \xi_n$
5. $(\xi_n)_{n=0}^{\infty}$ *is almost nowhere convergent; in fact, there is $\alpha > 0$ such that for almost all* $t \in [0,1]$ $\|\xi_n(t) - \xi_{n-1}(t)\| > \alpha/2$ *for all* $n \in \mathbb{N}$.

Remark 1.3: A uniformly bounded, simple-valued martingale $(\xi_n)_{n=0}^{\infty}$, where ξ_0 is a constant function, satisfying (5) is also called an $\alpha/2$-separated "bush" [J], a notation that will be used freely in the sequel.

What one usually does to obtain a bush is the following: Choose an arbitrary $x_\phi \in C$. The assumption, that the diameter of every slice of C has diameter greater than 2α quickly translates - via the Hahn-Banach theorem - to the fact that x_ϕ is contained in the closed, convex hull of $C \backslash B(x_\phi, \alpha)$, where $B(x_\phi, \alpha)$ denotes the ball of radius α around x_ϕ. So we may find $x_1, \ldots, x_{n_1}$ in C such that

$$\|x_k - x_\phi\| > \alpha \qquad k = 1, \ldots, n_1$$

and

$$\|n_1^{-1} \sum_{k=1}^{n_1} x_k - x_\phi\| \qquad \text{is very small.}$$

Repeating the argument we find for $k = 1, \ldots, n_1$ elements $x_{k,i} \in C$, $i = 1, \ldots, n_2$ such that

$$\|x_k - x_{k,i}\| > \alpha \qquad k = 1, \ldots, n_1, \quad i = 1, \ldots, n_2$$

and

$$\|n_2^{-1} \sum_{i=1}^{n_2} x_{k,i} - x_k\| \text{ very small for } \quad k = 1, \ldots, n_1$$

Continuing in an obvious way (or rather: obvious by now; it wasn't obvious 20 years ago) we obtain - up to small errors - an α-separated bush. Finally one gets rid of the errors by averaging back.

This bush may be considered as a simple-valued martingale $(\xi_n)_{n=0}^{\infty}$ as in the formulation of Th. 1.2. The additional ingredient stated by Th. 1.2 consists of the fact that - for the second generation of the bush - the differences $(x_{k,i} - x_k)$ above may be chosen to be in a space "almost orthogonal" to the space spanned by the x_k and so on (i.e., the differences between the bush-elements of the n'th generations and their predecessors may be chosen in a space "almost orthogonal" to the bush elements of the generations k, where $1 \leq k \leq (n-1)$).

How to achieve this extra property of the bush? Let E_1 be the subspace of X spanned by $\{x_1, \ldots, x_{n_1}\}$.

It is an old argument - due to S. Mazur (c.f. [L-T], Theorem 1.a.5) - that one may find a finite subset $\mathcal{E}_1 = \{x_1^*, \ldots, x_p^*\}$ in the unit-ball of X^* such that, for $z \in X$ such that $\langle z, x_j^* \rangle$ is small for each $1 \leq j \leq p$, z is "almost orthogonal" to E_1.

Suppose now that X fails (CPCP) which easily implies that we may find a closed convex C in the unit ball of X and $\alpha > 0$ such that every relatively weakly open subset V of C has diameter greater than 2α. For $k = 1, \ldots, n_1$ consider, for appropriate $\varepsilon_1 > 0$, the relatively weakly open sets

$$V_k\{x \in C : |\langle x, x_j^* \rangle| < \varepsilon_1 \quad \text{for} \quad 1 \leq j \leq p\} .$$

An application of the Hahn-Banach theorem furnishes, similarly as above, elements $x_{k,i} \in C$, such that the above conditions are satisfied, and in addition,

$$x_{k,i} - x_k \in V_k , \qquad \begin{array}{l} k = 1, \ldots, n_1 \\ i = 1, \ldots, n_2 \end{array}$$

i.e., the $(x_{k,i} - x_k)$ are "almost orthogonal" to E_1. With some additional care we may achieve that the space spanned by the differences $(x_{k,i} - x_k)$, $1 \leq k \leq n_1$, $1 \leq i \leq n_2$, is "almost orthogonal" to E_1. Continuing in a fairly obvious way one proves Theorem 1.2 for the case that X fails (CPCP) (for the details we refer to [B1] or [B2]).

The above argument led J. Bourgain to define the notion of (CPCP) as the proof of Th. 1.2 is now reduced to the case that X has (CPCP) but fails (RNP). It was only shown later by J. Bourgain and H.P. Rosenthal [B-R], that these notions (i.e., (RNP) and (CPCP)) do not coincide: In [B-R] the notion of (PCP) is defined, it is observed that (RNP) implies in fact (PCP) (which in turn obviously implies (CPCP)) and an example is given that (PCP) $\not\Rightarrow$ (RNP).

To apply (CPCP) to the present problem J. Bourgain proved the subsequent easy - but crucial - lemma, showing that (CPCP) implies (SR). The notion of strong regularity (used implicitly by J. Bourgain [B1] and later defined and developed by N. Ghoussoub, G. Godefroy and B. Maurey [G-G-M]) finally was the tool to complete the proof of Theorem 1.2.

Lemma 1.4 (see [B2] Lemma 5.3 or [G-G-M] Lemma II.1): *Let C be a convex, closed, bounded subset of a Banach space X and U a relatively weakly open subset of C. Then there are slices $S_1, \ldots, S_n$ of C s.t.*

$$n^{-1}(S_1 + \ldots + S_n) \subseteq U$$

Proof (c.f. B2]): Denote $\widetilde{C}$ the $\sigma(X^{**}, X^*)$-closure of C in X^{**} and fix $x \in U$. There is a $\sigma(X^{**}, X^*)$-neighbourhood V of x in X^{**} s.t. $(x + 2V) \cap C \subseteq U$. By the classical Krein-Milman-theorem (see, e.g., [D-U], p. 209) for compact sets there are extreme points $x_1^{**}, \ldots, x_n^{**}$ of $\widetilde{C}$ such that $n^{-1}(x_1^{**} + \ldots + x_n^{**})$ is in $x + V$. By the extremality of the x^* in $\widetilde{C}$ there are weak-star-slices $\widetilde{S}_k$ of $\widetilde{C}$ such that $\widetilde{S}_k \subseteq x_k^{**} + V$. Hence $S_k = \widetilde{S}_k \cap C$ are slices of C s.t.

$$\begin{aligned} n^{-1}(S_1 + \ldots + S_n) &\subseteq n^{-1} \sum_{k=1}^{n} (x_k^{**} + V) \cap C \\ &\subseteq (x + 2V) \cap C \subseteq U \,. \end{aligned}$$

□

Recently, it has been proved that all the notions of definition 1.1 are different: That (CPCP) $\not\Rightarrow$ (PCP) was shown in [G-G-M-S] and that (SR) $\not\Rightarrow$ (CPCP) in [A-O-R] and [G-M-S2].

We shall now prove Theorem 1.2 for the case that X fails (RNP) and has (CPCP), which is the "hard case" of the proof. Note that we shall in fact only use the fact that X fails to be strongly regular.

The technique of constructing a bush invented by J. Bourgain for this case will be completely different from the scheme of arguments used above.

Acknowledgement: We want to thank the referee for some essential improvements on the first version of this note.

2. The proof of Bourgain's theorem for the case that X has (CPCP).

Suppose that X has (CPCP) and fails (RNP) and fix - for the rest of the paper - a closed convex subset C of the unit ball of X and $\alpha > 0$ such that every slice of C has diameter greater than 2α.

Let $(\varepsilon_n)_{n=1}^{\infty}$ be a sequence in $]0,1]$ tending sufficiently fast to zero. To start the construction we do *not* choose an element x_ϕ as above but we choose slices $S_1, \ldots, S_{n_1}$ such that

$$\operatorname{diam}\big(n_1^{-1}(S_1 + \ldots + S_{n_1})\big) < \varepsilon_1 \ ,$$

which is possible by Lemma 1.4. The bush elements $x_1, \ldots, x_{n_1}$ will finally be chosen in $S_1, \ldots, S_{n_1}$ - up to a small perturbation. Note that whatever the choice of $x_k \in S_k$ will be, the average

$$x_\phi = n_1^{-1} \sum_{k=1}^{n_1} x_k$$

is already determined up to an ambiguity of at most ε_1.

Now make the crucial observation: If we make the S_k smaller, i.e., if we choose slices T_k contained in S_k, the above situation does not change in certain aspects: The diameter of each T_k will still have diameter greater than 2α while the diameter of the average $n_1^{-1}(T_1 + \ldots + T_{n_1})$ will be - a fortiori - less than ε_1. The art of the proof will consist in the proper choice of $T_k \subseteq S_k$. This idea was also exploited in [S].

The next lemma will be an easy but basic tool: Although the slices of C have diameter greater than 2α they may be chosen small in a local sense:

Lemma 2.1. ([B1], Lemma 3): *If S is a slice of C, $x_1^*, \ldots, x_p^* \in X^*$ and $\varepsilon > 0$ there is a slice T of C s.t.*

1. $\overline{T} \subseteq S$
2. oscillation $\{x_q^* \mid T\} \leq \varepsilon \qquad q = 1, \ldots, p.$

Proof: Denote again $\widetilde{C}$ the weak-star closure of C in the bidual. If C is of the form

$$S = S(x^*, \beta) = \{x \in C : \langle x, x^* \rangle > M_{x^*} - \beta\}$$

let

$$\widetilde{S} = \widetilde{S}(x^*, \beta) = \{x^{**} \in \widetilde{C} : \langle x^{**}, x^* \rangle > M_{x^*} - \beta\}$$

By Krein-Milman there is an extreme point x_0^{**} of $\widetilde{C}$ contained in $\widetilde{S}$. As

$$V = \{x^{**} \in \widetilde{S} : \langle x^{**} - x_0^{**}, x_q^* \rangle < \varepsilon/2 \ , \ q = 1, \ldots, p\}$$

is a weak-star-neighbourhood of x_0^{**} in $\widetilde{C}$ and as - by the separation theorem - the weak-star-slices of the extreme point x_0^{**} form a weak-star neighbourhood-basis of x_0^{**} in $\widetilde{C}$ there is

$$\widetilde{T} = \widetilde{T}(y^*, \gamma) = \{x^{**} \in \widetilde{C} : \langle x^{**}, y^* \rangle > M_{y^*} - \gamma\}$$

contained in V. The slice

$$T = T(y^*, \gamma/2) = \widetilde{T}(y^*, \gamma/2) \cap C$$

will satisfy 1. and 2. □

Back to working on the plan for the proof of Theorem 1.2: We have chosen the slices $S_1, \ldots, S_{n_1}$ and now want to make the clever choice of slices T_k in S_k. What we want to achieve is the following: We shall - later on - choose the elements $x_{k,i}$ $(k = 1, \ldots, n_1;\ i = 1, \ldots, n_2)$ of the second generation of the bush in T_k and obtain the elements of the first generation by averaging:

$$x_k = n_2^{-1} \sum_{i=1}^{n_2} x_{k,i} \ .$$

Note however, that at present we do not have any information, which $x_{k,i} \in T_k$ we shall choose; we don't even have a bound for n_2. Nevertheless, it will turn out that already now we can choose the $T_k \subseteq S_k$ in such a way that - whatever the choice of the $x_{k,i} \in T_k$ will be - the space E_1 spanned by x_k and the space E_2 spanned by the martingale differences $x_{k,i} - x_k$ will be - essentially, i.e., up to a certain perturbation (Lemma 2.7, below) - well complemented.

In order to obtain well-complemented finite-dimensional subspaces, we apply the old technique, due to Mazur, of constructing a basic sequence in a Banach space (c.f. [L-T], Theorem 1.a.5.):

Suppose for the moment we know already what the x_k's will be. Then there is a finite subset $\mathcal{E} = \{x_1^*, \ldots, x_p^*\}$ in the unit sphere of X^* which almost norms the space E_1 spanned by $\{x_1, \ldots, x_{n_1}\}$. Applying Lemma 2.1 we may find slices $T_k \subseteq S_k$ such that each element of $\mathcal{E}$ oscillates very little on each T_k. Hence the $x_{k,i} - x_k$ will be almost in the annihilator $\mathcal{E}^\perp$ of $\mathcal{E}$. A perturbation argument - which will need some additional care - will allow us to assume that the $x_{k,i} - x_k$ are in fact in $\mathcal{E}^\perp$. If we have achieved all that, it is well known - this is Mazur's argument - that the space E_2 spanned by the $x_{k,i} - x_k$ will be almost perpendicular to E_1, i.e., the projection from $\mathrm{span}(E_1, E_2)$ along E_2 onto E_1 has norm close to 1.

The next lemma will formulate this in more precise terms and point out a crucial observation: We don't really have to know the x_k's but only a finite set $\mathcal{E}$ which norms the space E_1

spanned by the x_k's, whatever the choice of the x_k will be. The reason for the appearance of F below will become clear later, when we have to deal with the perturbation argument.

Lemma 2.2: *Let F be a finite dimensional subspace of X, $T_1, \dots, T_n$ subsets of the unit ball of X, $\mathcal{E} = \{x_1^*, \dots, x_p^*\}$ a finite subset of the unit sphere of X^* and $1 > \varepsilon > o$.*

Suppose that for every choice of $x \in F$, $(a_k)_{k=1}^n \in \mathbb{R}^k$, $x_k \in T_k$ there is $x_q^* \in \mathcal{E}$ s.t.

$$(1+\varepsilon)\langle x + \sum_{k=1}^n a_k x_k \ , \ x_q^*\rangle \geq \|x + \sum_{k=1}^n a_k x_k\| \ .$$

Let $y \in \mathcal{E}^\perp$. Then for every choice of x, $(a_k)_{k=1}^n$ and $(x_k)_{k=1}^n$ as above

$$(1+\varepsilon)\|x + \sum_{k=1}^n a_k x_k + y\| \geq \|x + \sum_{k=1}^n a_k x_k\| \ .$$

Given any choice of $x_k \in T_k$ let E_1 be the space spanned by F and $\{x_1, \dots, x_n\}$ and E_2 a subspace of $\mathcal{E}^\perp$. Then the projection S from $\mathrm{span}(E_1, E_2)$ onto E_1 with kernel $(S) = E_2$ has norm less than $1 + \varepsilon$.

Proof: Obvious. □

Our program for the proof of Theorem 1.2 will therefore be almost completed, if we prove the next lemma (the space F will be defined later):

Lemma 2.3 ([B2], Lemma 5.10): *$S_1, \dots, S_n$ be slices of C, F a finite-dimensional subspace of X and $1 > \varepsilon > 0$. There are slices $T_k \subseteq S_k$ and a finite set $\mathcal{E} = \{x_1^*, \dots, x_p^*\}$ in the unit-sphere of X^* such that for every choice of $x \in F$, $(a_k)_{k=1}^n \in \mathbb{R}^n$, $(x_k)_{k=1}^n \in (T_k)_{k=1}^n$ we have*

$$(1+\varepsilon) \max_{1 \leq q \leq p} \langle x + \sum_{k=1}^n a_k x_k \ , \ x_q^*\rangle \geq \|x + \sum_{k=1}^n a_k x_k\| \ .$$

Remark 2.4: This lemma may be viewed as the core of the proof of Theorem 1.2. We shall need the following two sub-lemmata 2.5 and 2.6. The reasoning will be somewhat different from Bourgain's and similar to the proof of Lemma 2.7 of [S]. However, the difference is only superficial.

Sublemma 2.5: *Let $R_1 = S(x_1^*, \beta_1)$ be a slice of C, $1 > \gamma > 0$ and x^* in the unit-sphere of X^*.*

Denote $R_1' = S(x_1^*, \beta_1 \cdot \gamma/12)$ and suppose there is *some* $x_1 \in R_1'$ and $c_1 \in [-1, 1]$ such that

$$\langle x_1, x^* \rangle = c_1$$

Then there is a slice T_1 of C contained in R_1 such that for *every* $x_1 \in T_1$

$$\langle x_1, x^* \rangle > c_1 - \gamma/2$$

Proof: Applying again the Krein-Milman theorem we may find $(x_{1,j}^{**})_{j=1}^m$ in the weak-star-closure $\widetilde{C}$ of C in X^{**} such that we get

$$\langle m^{-1} \sum_{j=1}^m x_{1,j}^{**}, x^* \rangle > c_1 - \gamma/4$$

and

$$\langle m^{-1} \sum_{j=1}^m x_{1,j}^{**}, x_1^* \rangle > M_{x_1^*} - \beta_1 \cdot \gamma/12$$

On the other hand we know that, for $j = 1, \dots, m$

$$\langle x_{1,j}^{**}, x^* \rangle \leq 1$$

and

$$\langle x_{1,j}^{**}, x_1^* \rangle \leq M_{x_1^*} .$$

From elementary estimates we obtain

$$\#\{j : \langle x_{1,j}^{**}, x^* \rangle > c_1 - \gamma/2\} > m \cdot (\gamma/12)$$

and

$$\#\{j : \langle x_{1,j}^{**}, x_1^* \rangle > M_{x_1^*} - \beta_1\} > m \cdot (1 - \gamma/12) .$$

Hence there is an extreme-point x_{1,j_0}^{**} of $\widetilde{C}$ such that

$$\text{(a)} \quad \langle x_{1,j_0}^{**}, x^* > c_1 - \gamma/2$$

and

$$\text{(b)} \quad \langle x_{1,j_0}^{**}, x_1^* \rangle > M_{x_1^*} - \beta_1 .$$

Using again the fact that the weak-star-slices form a relative weak star neighbourhood-basis of x_{1,j_0}^{**} in $\widetilde{C}$, we may find a weak-star slice $\widetilde{T}_1$ of $\widetilde{C}$ such that (a) and (b) holds true for each $x^{**} \in \widetilde{T}_1$. The intersection $T_1 = \widetilde{T}_1 \cap C$ does the job. □

Sublemma 2.6: *Let* $R_k = S(x_k^*, \beta_k)$ *be slices of* C, $x_0 \in X$, x^* *in the unit-sphere of* X^*, $1 > \gamma > 0$ *and* $\vec{a} = (a_k)_{k=1}^n \in I\!R^n$ *such that*

$$\|\vec{a}\|_1 = \sum_{k=1}^n |a_k| \leq 1 .$$

Denote R'_k the slices $S(x_k^*, \beta_k \cdot \gamma/12)$ and suppose that there is $\tau \in \mathbb{R}$ such that for *some* choice $x_k \in R'_k$

$$\langle x_0 + \sum_{k=1}^{n} a_k x_k, x^* \rangle > \tau - \gamma/2 \ .$$

Then there are slices $T_k \subseteq R_k$ such that for *every* choice $x_k \in T_k$

$$\langle x_0 + \sum_{k=1}^{n} a_k x_k, x^* \rangle > \tau - \gamma \ .$$

Proof: For $k = 1, \ldots, n$ let

$$\langle x_k, x^* \rangle = c_k$$

and apply 2.5 to find $T_k \subseteq R_k$ such that for each $x_k \in T_k$

$$\langle x_k, \text{sign}(a_k) x^* \rangle > \text{sign}(a_k) c_k - \gamma/2 \ .$$

The desired conclusion then follows from $\|\vec{a}\|_1 \leq 1$. □

Proof of Lemma 2.3: Let G denote the finite dimensional space $F \oplus_1 \ell_n^1$ and choose a γ-net $(z_q, \vec{a}_q)_{q=1}^p = \left(z_q, (a_{q,i})_{i=1}^n\right)_{q=1}^p$ in the unit sphere of G, where $\gamma < \varepsilon\alpha/36n$. We proceed by induction on $q = 1, \ldots, p$. For $q = 1$ consider

$$\tau(z_1, \vec{a}_1) = \inf\{\sup \|x_1 + \sum_{k=1}^{n} a_{1,k} x_k\| : x_k \in R_k\}$$

the inf taken over all slices R_k contained in S_k. Find slices $R_k \subseteq S_k$ such that the above infimum is approached up to γ. If $R_k = S(x_k^*, \beta_k)$ let $R'_k = S(x_k^*, \beta_k \gamma/12)$ and find x_1^* in the unit sphere of X^* such that

$$\sup\{\langle x_1^*, z_1 + \sum_{k=1}^{n} a_{1,k} x_k \rangle : x_k \in R'_k\} > \tau(z_1, \vec{a}_1) - \gamma/2 \ .$$

By Sublemma 2.6 we may find slices $T_k^1 \subseteq R_k$ such that for each choice of $x_k \in T_k^1$

$$\langle x_1^*, z_1 + \sum_{k=1}^{n} a_{1,k} x_k \rangle > \tau(z_1, \vec{a}_1) - \gamma \ .$$

On the other hand, by the construction of the R_k we get for each choice of $x_k \in T_k^1$

$$\|z_1 + \sum_{k=1}^{n} a_{1,k} x_k\| < \tau(z_1, \vec{a}_1) + \gamma \ .$$

Let $S_k^1 = T_k^1$ and repeat the argument with $(z_1, \vec{a}_1)$ replaced by $(z_2, \vec{a}_2)$ and S_k by S_k^1. Continuing in an obvious way we obtain slices $S_k \supseteq S_k^1 \supseteq \cdots \supseteq S_k^{p-1} \supseteq T_k^p = T_k$ such that for every $q = 1, \ldots, p$ and for every choice $x_k \in T_k^q$ (hence in particular for $x_k \in T_k$)

$$\langle x_q^*, z_q + \sum_{k=1}^{n} a_{q,k} x_k \rangle > \tau(z_q, \vec{a}) - \gamma \tag{1}$$

while

$$\| z_q + \sum_{k=1}^{n} a_{q,k} x_k \| < \tau(z_q, \vec{a}_q) + \gamma \;, \tag{2}$$

where the $\tau(z_q, \vec{a}_q)$ we define relative to the slices $(S_k^{q-1})_{k=1}^n$. We need the following rough estimate showing that, for $q = 1, \ldots, p$, the figure $\tau(z_q, \vec{a}_1)$ is big compared to γ:

$$\tau(z_q, \vec{a}_q) \geq \alpha / 3n \;. \tag{3}$$

Indeed, if $\|z_q\| \geq 2/3$, then for every choice of slices $R_k \subseteq S_k^{q-1}$ and every choice $x_k \in R_k$

$$\| z_q + \sum_{k=1}^{n} a_{q,k} x_k \| \geq 2/3 - 1/3 = 1/3 \;.$$

If $\|z_q\| < 2/3$ then there is k_0 such that $|a_{q,k_0}| \geq 1/3n$. Given slices $R_k \subseteq S_k^{q-1}$ and given any choice of $x_k \in R_k$ for $k \neq k_0$ we infer from the fact that the diameter of R_{k_0} is greater than 2α that we can always make a choice $x_{k_0} \in R_{k_0}$ such that

$$\| z_q + \sum_{k=1}^{n} a_{q,k} x_k \| \geq \alpha / 3n$$

which shows (3).

Consider now an arbitrary $(z, \vec{a})$ in the unit sphere of G and find q such that $\|(z, \vec{a}) - (z_q, \vec{a}_q)\|_G < \gamma$. For every choice $x_k \in T_k$ we get

$$\langle x_q^*, z + \sum_{k=1}^{n} a_k x_k \rangle \geq \langle x_q^*, z_q + \sum_{k=1}^{n} a_{q,k} x_k \rangle - \gamma \geq \tag{4}$$

$$\geq \| z_q + \sum_{k=1}^{n} a_{q,k} x_k \| - 3\gamma \geq$$

$$\geq \| z + \sum_{k=1}^{n} a_k x_k \| - 4\gamma \;. \tag{5}$$

and from (1), (3) and (4) above we conclude that

$$\langle x_q^*, z + \sum_{k=1}^{n} a_k x_k \rangle \geq \tau(z_q, \vec{a}_q) - 2\gamma$$

$$\geq \alpha/3n - 2\gamma \ . \tag{6}$$

Multiplying (6) with ε and adding to (5) we conclude from $\gamma < \varepsilon\alpha/36n$ that

$$(1+\varepsilon) \max_{q=1,\ldots,p} \langle x_q^*, z + \sum_{k=1}^{n} a_k x_k \rangle \geq \|z + \sum_{k=1}^{n} a_k x_k\| \tag{7}$$

which holds for every choice $x_k \in T_k$ and for every $(z, \vec{a})$ in the unit sphere of G. As (7) is homogeneous it holds true for every $(z, \vec{a}) \in G$ thus proving Lemma 2.3. □

Let us come back to the strategy of proving Theorem 1.2 as discussed after Lemma 2.1. We have now assembled almost all ingredients for the proof except the "perturbation argument" announced there: What we can do so far is only to ensure that for any choice of elements $x_{k,i} \in T_k$, $k = 1, \ldots, n_1$, $i = 1, \ldots, n_2$ the martingale differences $x_{k,i} - x_i$ are close to the annihilator $\mathcal{E}^{\perp}$. We shall replace $x_{k,i}$ by $y_{k,i}$ with $\|x_{k,i} - y_{k,i}\|$ small and such that for

$$y_k = n_2^{-1} \sum_{i=1}^{n_2} y_{k,i}$$

we obtain that $y_{k,i} - y_k$ is in $\mathcal{E}^{\perp}$. However, there arises a new difficulty: Although the y_k will of course be close to the x_k, there is no reason that the space spanned by the y_k is close to the space spanned by the x_k. More precisely: If Lemma 2.3 tells us that we may find $\mathcal{E}$ that almost norms the span of x_k, whatever the choice $x_k \in T_k$ will be, we have no way to ensure that $\mathcal{E}$ will almost norm the space spanned by y_k, if we only assume that $\|x_k - y_k\|$ is small.

The following elementary lemma will allow us to choose the perturbations $x_{k,i} - y_{k,i}$ not only small in norm but lying in prefixed finite dimensional space. The observation (b) will make life easier later (with respect to the technicalities of the proof).

Lemma 2.7 (c.f.[B1], Lemma 18): *Let $\mathcal{E} = \{x_1^*, \ldots, x_p^*\}$ in X^* and $\varepsilon > 0$. There is $\delta > 0$ and a finite dimensional subspace F of X such that*

(a) *For $x \in X$ with $|\langle x, x_q^* \rangle| \leq \delta$ for $q = 1, \ldots, p$ there is $y \in F$ with $\|y\| < \varepsilon/2$ and $\langle x_q^*, x \rangle = \langle x_q^*, y \rangle$ for $q = 1, \ldots, p$.*

(b) *If* $x_1, \ldots, x_n \in X$ *with* $n^{-1}\sum_{i=1}^n x_i = 0$ *and* $|\langle x_i, x_q^*\rangle| \le \delta$ *for* $q = 1, \ldots, p$, $i = 1, \ldots, n$ *there are* $y_1, \ldots, y_n \in F$ *with* $\|y_i\| < \varepsilon$, $\langle x_q^*, x_i\rangle = \langle x_q^*, y_i\rangle$ *for* $q = 1, \ldots, p$ *and*

$$n^{-1}\sum_{i=1}^{n} y_i = 0 \ .$$

Proof: Part (a) is an elementary exercise (see [B1], Lemma 18). For part (b) apply part (a) to obtain $w_1, \ldots, w_n$ in F, $\|w_i\| < \varepsilon/2$ and

$$\langle x_q^*, w_i\rangle = \langle x_q^*, x_i\rangle \qquad q = 1, \ldots, p \ , \ i = 1, \ldots, n$$

Let $w = n^{-1}\sum_{i=1}^n w_i$ and note that for $q = 1, \ldots, p$

$$\langle x_q^*, w\rangle = \langle x_q^*, n^{-1}\sum_{i=1}^{n} x_i\rangle = 0 \ .$$

Hence $y_i = w_i - w$ are elements of F, $\|y_i\| < \varepsilon$ and satisfy (b). □

We now have assembled all the ingredients for the first step of the construction of the bush described in Theorem 1.2, namely we have furnished all the machinery to make the proper choice of slices $T_k \subseteq S_k$ for $k = 1, \ldots, n_1$.

Suppose now we have fixed the T_k's of the first generation.

For the second step we shall choose slices $S_{k,i} \subseteq T_k$, where $i = 1, \ldots, n_2$, such that

$$\text{diameter}\left(n_2^{-1}\sum_{i=1}^{n_2} S_{k,i}\right) < \varepsilon_2 \ . \qquad k = 1, \ldots, n_1$$

The assertion that this is possible under the assumption (CPCP) (in fact, (SR) would suffice) as well as the fact guaranteeing that the bush $\{x_\omega : \omega \in \Omega\}$ will be $\alpha/2$-separated will be furnished by the subsequent lemma 2.8.

The bush-elements $x_{k,i}$ of the second generation will eventually be chosen in $S_{k,i}$ (hence in particular they will be in T_k).

Lemma 2.8: *Let* S *be a slice of* C *and* $\varepsilon \in]0, \alpha/8[$. *There are slices* $(S_i)_{i=1}^n$ *contained in* S *such that*

$$\text{(i)} \quad \text{diam}\left(n^{-1}\sum_{i=1}^{n} S_i\right) < \varepsilon$$

$$\text{(ii)} \quad \text{distance}\left(n^{-1}\sum_{j=1}^{n} S_j, S_i\right) > 3\alpha/4 \quad \text{for } 1 \le i \le n \ .$$

Proof: If $S = S(x^*, \beta)$ let $S' = S(x^*, \beta/2)$. It follows from the assumption that C has CPCP that S' contains a relatively weakly open subset U of diameter less than $\varepsilon/2$. Note that U is relatively weakly open in C too, hence by Lemma 1.4 we may find finitely many slices $R_1, \ldots, R_m$ of C such that

$$m^{-1} \sum_{j=1}^{m} R_j \subseteq U .$$

Note that we can not assure that the R_j are contained in S. However by a reasoning similar to the proof of Lemma 2.5 we conclude from $U \subseteq S'$ that at least half of the $(R_j)_{j=1}^m$ are contained in S. Relabeling those R_j by $T_1, \ldots, T_n$ we obtain slices of C contained in S and such that

$$\text{diameter}\left(n^{-1} \sum_{i=1}^{n} T_i\right) < \varepsilon .$$

To obtain (ii) fix $i \in \{1, \ldots, n\}$ and let z be an arbitrary element of $n^{-1} \sum_{i=1}^n T_i$. If $T_i = S(x_i^*, \beta_i)$ let $T_i' = S(x_i^*, \beta_i \cdot \alpha/24)$. As diameter$(T_i') > 2\alpha$ we may find $z_i \in T_i'$ such that

$$\|z - z_i\| > \alpha .$$

Find $x^* \in X^*$, $\|x^*\| = 1$ such that

$$\langle z_i - z, x^* \rangle > \alpha$$

and apply Sublemma 2.5 with $c_i = \langle z_i, x^* \rangle$ to find a slice $S_i \subseteq T_i$ such that for each $x_i \in T_i$

$$\langle x_i, x^* \rangle > c_i - \alpha/8$$

hence

$$\langle x_i - z, x^* \rangle > 7\alpha/8 .$$

From the assumption $\varepsilon < \alpha/8$ and (i) we obtain for $x_i \in S_i$ and $x \in n^{-1} \sum_{i=1}^n T_i$

$$\|x - x_i\| > 3\alpha/4 .$$

Doing this procedure successively for $i = 1, \ldots, n$ we have also proved (ii). □

We are now in the position that we have all the tools for the final proof.

3. Proof of Theorem 1.2: Fix a sequence $(\varepsilon_p)_{p=1}^\infty$ in $]0,1]$ such that $\prod_{p=1}^{\infty} (1 + \varepsilon_p) < \lambda$ and $\sum_{p=1}^{\infty} \varepsilon_p < \alpha/8$. We shall proceed by induction on p:

For $p = 1$ apply Lemma 2.8 to $S = C$ and $\varepsilon = \varepsilon_1$ fo find $n_1 \in I\!N$ and slices $(S_{i_1})_{i_1=1}^{n_1}$ such that 2.8 holds true.

Apply Lemma 2.3 with $F = \{0\}$ and $\varepsilon = \varepsilon_1$ to find a finite subset $\mathcal{E}_1$ of the unit sphere of X^* and slices $R_{i_1} \subseteq S_{i_1}$ such that 2.3 holds.

Now apply Lemma 2.7 to $\mathcal{E}_1$ and ε_1 to find a finite dimensional $F_1 \subseteq X$ and $\delta_1 > 0$ satisfying 2.7.

Finally apply Lemma 2.1 to find slices $T_{i_1} \subseteq \overline{T}_{i_1} \subseteq R_{i_1}$ such that each $x^* \in \mathcal{E}_1$ oscillates on each T_{i_1} less than δ_1.

For the general induction step suppose that there have been chosen finite dimensional subspaces $F_1 \subseteq \ldots \subseteq F_{p-1}$ of X, finite subsets $\mathcal{E}_1 \subseteq \mathcal{E}_2 \subseteq \ldots \subseteq \mathcal{E}_{p-1}$ of the unit sphere of X^*, natural numbers $n_1, \ldots, n_{p-1}$ and for every $(i_1, \ldots, i_{p-1}) \in \{1, \ldots, n_1\} \times \{1, \ldots, n_2\} \times \{1, \ldots, n_{p-1}\} = \Omega_{p-1}$ slices $T_{i_1}, \ldots, i_{p-1}$.

Apply Lemma 2.8 to find $n_p \in I\!N$ and slices $S_{i_1,\ldots,i_p}$ indexed by $\Omega_p = \{1, \ldots, n_1\} \times \ldots \times \{1, \ldots, n_p\}$ such that for each $\{i_1, \ldots, i_{p-1}\} \in \Omega_{p-1}$ and $1 \leq i_p \leq n_p$

$$- \quad S_{i_1,\ldots,i_p} \subseteq T_{i_1,\ldots,i_{p-1}}$$

$$(8) \qquad - \quad \operatorname{diam}\left(n_p^{-1} \sum_{i_p=1}^{n_p} S_{i_1,\ldots,i_p}\right) < \varepsilon_p$$

$$- \quad \text{for} \quad x \in n_p^{-1} \sum_{i_p=1}^{n_p} S_{i_1,\ldots,i_p} \quad \text{and}$$

$$(9) \qquad y \in S_{i_1,\ldots,i_p} \quad \text{we have} \quad \|x - y\| > 3\alpha/4\ .$$

It is easy to convince oneself that one may choose one single $n_p \in I\!N$ working for all $(i_1, \ldots, i_{p-1})$ in the finite set Ω_{p-1} simultaneously.

Now apply Lemma 2.2 to $F = F_{p-1}$, $\varepsilon = \varepsilon_p$ and the collection of slices $S_{i_1,\ldots,i_p}$ to find a finite subset $\mathcal{E}_p \supseteq \mathcal{E}_{p-1}$ of the unit sphere of X^*, slices $R_{i_1,\ldots,i_p} \subseteq S_{i_1,\ldots,i_p}$ such that 2.2 holds.

Next apply Lemma 2.7 to $\mathcal{E}_p$ and ε_p to find a finite dimensional $F_p \supseteq F_{p-1}$ and $\delta_p > 0$ satisfying 2.7.

Finally apply Lemma 2.1 to find slices $T_{i_1,\ldots,i_p} \subseteq \overline{T}_{i_1,\ldots,i_p} \subseteq S_{i_1,\ldots,i_p}$ such that each $x^* \in \mathcal{E}_p$ oscillates on each $T_{i_1,\ldots,i_p}$ less than S_p.

This finishes the induction step.

We now pass from the "bush of slices" to a bush of points $\mathcal{X}$ by means of (8) and the "averaging-back-technique": For each $(i_1,\dots,i_p)\in\Omega_p$ choose an arbitary

$$u_{i_1,\dots,i_p}\in T_{i_1,\dots,i_p}\ .$$

Define

$$x_{i_1,\dots,i_p}=\lim_{1\mapsto\infty} n_{p+1}^{-1}\cdot\ldots\cdot n_q^{-1}\sum_{i_{p+1}=1}^{n_{p+1}}\cdots\sum_{i_q=1}^{n_q}u_{i_1,\dots,i_p,\dots,i_q}\ .$$

It follows from (8) that the limit exists; it is, of course, contained in $\overline{T}_{i_1,\dots,i_p}$ and we infer from (9) that

$$\|x_{i_1,\dots,i_p}-x_{i_1,\dots,i_p,i_{p+1}}\|\geq 3\alpha/4\ .$$

Clearly the $x_{i_1,\dots,i_p}$ form a bush as each element is the arithmetic mean of its successors.

To be able to apply the conclusion of Lemma 2.3 we still have to apply the perturbation argument. For $p\geq 1$ and $(i_1,\dots,i_p)\in\Omega_p$ and $1\leq i_{p+1}\leq n_{p+1}$ note that for $x^*\in\mathcal{E}_p$

$$\|\langle x_{i_1,\dots,i_{p+1}}-x_{i_1,\dots,i_p},x^*\rangle|\leq\delta_p\ .$$

Hence we may apply Lemma 2.7(b) to find $w_{i_1,\dots,i_{p+1}}\in F_p$ with

– $\|w_{i_1,\dots,i_{p+1}}\|<\varepsilon_p$

– $\langle x_{i_1,\dots,i_{p+1}}-x_{i_1,\dots,i_p},x^*\rangle=\langle w_{i_1,\dots,i_{p+1}},x^*\rangle$ for $x^*\in\mathcal{E}_p$

(10) – $\displaystyle\sum_{i_{p+1}=1}^{n_{p+1}}w_{i_1,\dots,i_{p+1}}=0$

We define a "bush of errors " $\mathcal{Z}$ on the index set $\bigcup_{p=1}^{\infty}\Omega_p$ inductively by

$$z_{i_1}=z_{i_2}=\ldots=z_{i_p}=0$$

and for $p\geq 1$

$$z_{i_1,\dots,i_{p+1}}=z_{i_1,\dots,i_p}+w_{i_1,\dots,i_{p+1}}\ .$$

It follows from (10) that $\mathcal{Z}$ is indeed a bush and we get the estimate

$$\|z_{i_1,\dots,i_{p+1}}\|\leq\sum_{q-1}^{p}\varepsilon_q<\alpha/8\ .$$

The bush $\mathcal{Y}$ is obtained by subtracting $\mathcal{Z}$ from $\mathcal{X}$: Letting

$$y_{i_1,\dots,i_p}=x_{i_1,\dots,i_p}-z_{i_1,\dots,i_p}$$

this defines a bounded bush in X with

$$\|y_{i_1,\dots,i_{p+1}} - y_{i_1,\dots,i_p}\| \geq \|x_{i_1,\dots,i_{p+1}} - x_{i_1,\dots,i_p}\| - 2\cdot\alpha/8 \geq \alpha/2$$

Let $E_1 = \text{span}\{y_{i_1}, \dots, y_{i_{n_1}}\}$ and, for $p \geq 2$

$$E_p = \text{span}\{y_{i_1,\dots,i_p} - y_{i_1,\dots,i_{p-1}} : (i_1,\dots,i_{p-1}) \in \Omega_{p-1}\ ,\ 1 \leq i_p \leq n_p\}$$

We claim that $(E_p)_{p=1}^{\infty}$ is a finite-dimensional Schauder-decomposition of its span with Grynblum-constant less than λ. Indeed, for $p \in I\!N$ let

$$F_p = \text{span}(E_1, \dots, E_p)$$

$$\text{and} \qquad G_p = \overline{\text{span}}\Big(\bigcup_{q=p+1}^{\infty} E_q\Big)\ .$$

It will suffice to show that $E = F_p \oplus G_p$ and that the projection $S_p : E \mapsto F_p$ with kernel $(S_p) = G_p$ satisfies $\|S_p\| < (1 + \varepsilon_p)$.

To see this, note that F_p is contained in the span of the $x_{i_1,\dots,i_q}$ and $z_{i_1,\dots,i_q}$ with $q \leq p$. Also note that these $x_{i_1,\dots,i_q}$ are contained in $\overline{T}_{i_1,\dots,i_q} \subseteq S_{i_1,\dots,i_q}$ and the $z_{i_1,\dots,i_q}$ are contained in F_{p-1} (for $q \leq p$).

On the other hand G_p is contained in $\mathcal{E}_p^{\perp}$ the annihilator of $\mathcal{E}_p$. So we are exactly in the situation of Lemma 2.2 to conclude that $\|S_p\| \leq 1 + \varepsilon_p$. This finishes the proof of Theorem 1.2.

□

References

[A-O-R] S. Argyros, E. Odell, H.P. Rosenthal. In preparation

[B1] J. Bourgain. Dentability and finite-dimensional decompositions. Studia Math. 67 (1980), 135-148.

[B2] J. Bourgain. La propriété de Radon-Nikodym. Publications Mathématiques de l'Université Pierre et Marie Curie, no. 36, (1979).

[B-R] J. Bourgain, H.P. Rosenthal. Geometrical implications of certain finite-dimensional decompositions. Bull. Soc. Math. Belg. 32 (1980), 57-82.

[D-U] J. Diestel, J.J. Uhl. Vector measures. AMS, Providence, 1977.

[E-W] G.A. Edgar, R. Wheeler. Topological properties of Banach spaces. Pac. J. of Math. 115 (1984), 317-350.

[G-G-M] N. Ghoussoub, G. Godefroy, B. Maurey. First class functions around a subset of a compact space and geometrically regular Banach spaces. Preprint.

G-G-M-S] N. Ghoussoub, G. Godefroy, B. Maurey, W. Schachermayer. Some topological and geometrical structures in Banach spaces (essentially containing [G-G-M]), (1986), to appear.

[G-M-S1] N. Ghoussoub, B. Maurey, W. Schachermayer. A counter-example to a problem about points of continuity in Banach spaces. (1985), to appear in Proc. of AMS.

[G-M-S2] N. Ghoussoub, B. Maurey, W. Schachermayer. Geometrical implications of certain infinite-dimensional decompositions. In preparation.

[J] R.C. James. Subbushes and extreme points in Banach spaces. In "Proceedings of a Research Workshop in Bach Spaces Theory", University of Iowa, editor: Bor-Luh-Lin, (1981), 59-81.

[L-T] J. Lindenstrauss, L. Tzafriri. Classical Banach Spaces I. Springer (1977).

[S] W. Schachermayer. The Radon-Nikodym property and the Krein-Milman property are equivalent for strongly regular sets. (1986), to appear in Transactions of the AMS.

SUDAKOV TYPE INEQUALITIES FOR CONVEX BODIES IN $\mathbb{R}^n$

V.D. Milman* and **N. Tomczak-Jaegermann**

School of Mathematical Sciences
Raymond and Beverley Sackler
Faculty of Exact Sciences
Tel Aviv University
Tel Aviv, Israel

Department of Mathematics
University of Alberta
Edmonton, Alberta
Canada T6G 2E1

1. Consider $\mathbb{R}^n$ equipped with the natural Euclidean structure, i.e. with the inner product denoted by $(\cdot,\cdot)$ and the Euclidean norm $\|x\|_2^2 = (x,x)$. Let $D_n = \{x \in \mathbb{R}^n\colon \|x\| \le 1\}$ be the unit Euclidean ball and let $K \subset \mathbb{R}^n$ be a compact central symmetric convex body.

Recall, that the *covering number* $N(K,T)$ of the convex body K by another convex body T is

$$N(K,T) = inf\{N | \exists \{x_i\}_{i=1}^N \subset \mathbb{R}^n \text{ such that } K \subset \cup(x_i + T)\}.$$

We are interested in this note to estimate the covering numbers $N(K,\varepsilon D)$ and $N(D,\varepsilon K)$ in the spirit of the Sudakov minoration estimate. To introduce this estimate we need to develop some language.

Let $\|\cdot\|_K$ be the norm in $\mathbb{R}^n$ such that K is the unit ball in this norm. Define

$$M(K) = \int_{S^{n-1}} \|x\|_K d\mu(x)$$

where $S^{n-1} = \partial D_n$ is the unit Euclidean sphere and μ is the normalized rotation invariant measure on S^{n-1}. The polar body $K^\circ = \{x \in \mathbb{R}^n | \mid (x,y) \mid \le 1 \text{ for all } y \in K\}$. In particular,

$$M(K^\circ) = \int_{S^{n-1}} \sup_{y\in K} |(x,y)| d\mu(x) = \frac{c_n}{\sqrt{n}} \int_{\mathbb{R}^n} \sup_{y \in K} |(x,y)| d\gamma_n(x),$$

where γ_n is the standard Gaussian probability measure on $\mathbb{R}^n$ and $0 < \inf_n c_n < \sup_n c_n < \infty$ (cf. [T], II.5).

Theorem A (Sudakov, [Su]) Let $K \subset \mathbb{R}^n$ be a compact convex body. Then

$$N(K,D) \le \exp(cnM(K^\circ)^2),$$

* Supported in part by the Fund for Basic Research administered by the Israel Academy of Sciences and Humanities

where $c \geq 1$ is a universal constant. In particular, for every $\varepsilon \geq 0$,

$$N(K, \varepsilon D) \leq \exp(cn(M(K^\circ)/\varepsilon)^2).$$

We will also use the "dual" statement to the above estimate which follows directly from [PT1] and [C] (cf. also [PT2], Th. A and 1)*. It states

Theorem B. $N(D,K) \leq exp(cnM(K)^2)$, where $c \geq 1$ is a universal constant.

We show below how to improve both estimations (Theorems A and B) under some natural conditions on the behaviour of the covering numbers as functions of ε. Moreover, these new bounds on $N(K, \varepsilon D)$ and $N(D, \varepsilon K)$ are already, under some conditions, exact. We describe these conditions in terms of mixed volumes $V_k(K), k = 0, 1, \ldots, n$, of a convex body $K \subset \mathbb{R}^n$ and the n-dimensional ball $D(= D_n)$.

Let Vol_n denote a standard Lebesgue measure on $\mathbb{R}^n$ and let Vol_k be the k-dimensional Lebesgue measure on a k-dimensional subspace E of $\mathbb{R}^n$. Also let P_E be the orthogonal projection of $\mathbb{R}^n$ onto E. Then, recall, (see, e.g. [BZ] or [S])

$$V_k(K) = \frac{\mathrm{Vol}_n D_n}{\mathrm{Vol}_k D_k} \int_{E \in G_{n,k}} \mathrm{Vol}_k(P_E K) d\mu_k(E)$$

where μ_k is the rotation invariant (Haar) probability measure on the Grassman manifold $G_{n,k}$ (of all k-dimensional subspaces E of $\mathbb{R}^n$) - see, e.g., [MSch]. Note that $V_n(K) = \mathrm{Vol}_n K$ and, by the definition $V_0(K) = \mathrm{Vol}_n D_n$. We mostly deal with other quantities (for $k = 1, \ldots, n$):

$$A_k(K) = [V_k(K)/\mathrm{Vol}_n(D_n)]^{1/k}$$

which are compared by the following *Alexandroff inequalities* (see [B.-Z.]).

$$A_n(K) \leq A_{n-1}(K) \leq \cdots \leq A_1(K).$$

Note, that $A_1(K) = M(K^\circ)$.

We start our construction. For $K \subset \mathbb{R}^n$, as before a convex symmetric compact body, and $t \geq 0$ define the body

$$K_t = K \cap tD.$$

---

* A. Pajor has shown us a direct proof of theorem B without use of any non-trivial results.

Then $\|x\|_{K_t} = \max(\|x\|_K, t^{-1}\|x\|_2)$, for $x \in I\!R^n$. Write $K_t^\circ = (K_t)^\circ$ and set

$$M_t = M(K_t^\circ) = M(\mathrm{conv}(K^\circ \cup \frac{1}{t}D)).$$

Clearly, M_t is a non-decreasing function of t and M_t/t is a non-increasing function. Also, $M_t = t$ for t sufficiently small and $M_t = M(K^\circ)$ for t sufficiently large.

It is well known, (see, e.g. [MSch]) that, with "high" probability, k-dimensional sections of K_t° are "close" to euclidean k-dimensional balls if $k \underset{\sim}{<} n(M_t/t)^2$. Precisely:

Theorem C. Under the above notations, for every $0 < \delta < 1$ and $C > 1$ there exists $\eta = \eta(\delta; C) > 0$ such that for every $k \le k_0 = [\eta \cdot n(M_t/t)^2]$ there exists a subset $\mathcal{A} \subset G_{n,k}$ of k-dimensional subspaces of $I\!R^n$ with $\mu_k(\mathcal{A}) \ge 1 - e^{-Ck}$ and, for every $E \in \mathcal{A}$,

$$(1-\delta)\frac{1}{M_t}(D \cap E) \subset (K_t^\circ \cap E) \subset (1+\delta)\frac{1}{M_t}(D \cap E).$$

In the dual version

$$(1+\delta)^{-1}M_t(P_E D) \subset P_E K_t \subset (1-\delta)^{-1}M_t(P_E D).$$

Below, we will often use the following easy observation:

let D be an euclidean ball and K be a convex set; for any $y \in I\!R^n$ there exists $x \in K$ such that

$$K \cap (D+y) \subset K \cap (D+x) .$$

2. Theorem 1. a) Let $s > 1$ satisfy

(1) $$N(K, s\varepsilon D)^2 \le N(K, \varepsilon D).$$

Then

$$N(K, \varepsilon D) \le \exp(Cns^2(M_\varepsilon/\varepsilon)^2)$$

where C is a universal constant.

b) For every $\beta > 0$ and an integer k the condition

$$A_k(K)/M_\varepsilon \ge 2(1+\beta),$$

implies

$$\exp(\beta k/2) \le N(K, \varepsilon D).$$

Proof. The proof of the upper estimate a) is based on the following submultiplicative inequality, valid for all $\varepsilon > 0$ and $\tau > 0$:

$$N(K, \varepsilon D) \leq N(K \cap \frac{\tau}{2} D, \frac{\varepsilon}{2} D) \cdot N(K, \tau D). \tag{2}$$

Assuming the truth of (2) and setting $\tau = s\varepsilon$, we get, by (1),

$$\begin{aligned} N(K, \varepsilon D) &\leq N(K \cap \frac{s\varepsilon}{2} D, \frac{\varepsilon}{2} D) N(K, s\varepsilon D) \\ &\leq N(K_{s\varepsilon/2}, \frac{\varepsilon}{2} D)\, N(K, \varepsilon D)^{\frac{1}{2}}. \end{aligned}$$

Hence, by Theorem A,

$$\begin{aligned} N(K, \varepsilon D) &\leq N(K_{s\varepsilon/2}, \frac{\varepsilon}{2} D)^2 \\ &\leq \exp(Cn(M_{s\varepsilon/2}/\varepsilon)^2) \\ &\leq \exp(Cns^2(M_\varepsilon/\varepsilon)^2) . \end{aligned}$$

To prove (2), set $N_1 = N(K, \tau D)$ and $N_2 = N(K \cap \frac{\tau}{2} D, \frac{\varepsilon}{2} D)$. There exist $x_1, \ldots, x_{N_1}$ in K such that

$$K \subset \bigcup_{i=1}^{N_1} (x_i + \tau D) \cap K. \tag{3}$$

Consider $z_1, \ldots, z_{N_2}$ in $\mathbb{R}^n$ such that

$$K \cap \frac{\tau}{2} D \subset \bigcup_{m=1}^{N_2} (z_m + \frac{\varepsilon}{2} D).$$

It is easy to see that for every $i = 1, \ldots, N_1$,

$$(x_i + \tau D) \cap K \subset \bigcup_{m=1}^{N_2} (x_i + 2z_m + \varepsilon D).$$

Combining this covering with (3) we obtain $N(K, \varepsilon D) \leq N_1 N_2$, which concludes the proof of (2).

□

To prove b), note that $K \subset \bigcup_1^N (x_i + \varepsilon D)$ implies

$$K \subset \bigcup_1^N (x_i + \varepsilon D) \cap K \subset \bigcup_1^N (x_i + 2(K \cap \varepsilon D)). \tag{4}$$

For every $E \in G_{n,k}$, take P_E; then

$$\mathrm{Vol}_k P_E K \le N \mathrm{Vol}_k 2 P_E(K \cap \varepsilon D).$$

Integrating by $E \in G_{n,k}$ we have

$$N(K, \varepsilon D) \ge [A_k(K)/2A_k(K_\varepsilon)]^k$$

By Alexandroff inequality $M_\varepsilon = M(K_\varepsilon^\circ) = A_1(K_\varepsilon) \ge A_k(K_\varepsilon)$. Therefore, $N(K, \varepsilon D) \ge (A_k(K)/2M_\varepsilon)^k \ge \exp(\beta k/2)$.

Remarks 1. By extension of Brunn Theorem (see, e.g., [Ch], Lemma 2) $\mathrm{Vol}((x + \varepsilon D) \cap K) \le \mathrm{Vol}(\varepsilon D \cap K)$, for any central symmetric convex body K. Therefore, we have straightforward from (4) another lower bound for the covering number:

$$N(K, \varepsilon D) \ge \mathrm{Vol} K / \mathrm{Vol}(K \cap \varepsilon D).$$

2. Alexandroff's inequality $M_\varepsilon \ge A_k(K_\varepsilon)$ used in the proof, is, for some k, exact. By Theorem C, for $k < \eta n (M_\varepsilon/\varepsilon)^2$ we have an equivalence $M_\varepsilon \simeq A_k(K_\varepsilon)$. More precisely, for every $\delta > 0$ there exists $\eta = \eta(\delta) > 0$ such that for any integer $k < \eta \cdot n \cdot (M_\varepsilon/\varepsilon)^2$,

$$1 \le M_\varepsilon / A_k(K_\varepsilon) \le 1 + \delta.$$

3. Obviously, $A_k(K) \ge A_k(K_{t\varepsilon})$ for any $t > 0$. Therefore, using the previous Remark, we obtain that $A_k(K_{t\varepsilon}) \ge M_{t\varepsilon}/(1+\delta)$ for $k \le \eta \cdot n \cdot (M_{t\varepsilon}/t\varepsilon)^2$ and

$$N(K; \varepsilon D) \ge (M_{t\varepsilon}/2(1+\delta)M_\varepsilon)^k.$$

Combining Theorem 1 and Remark 3, we derive the following.

Corollary 2. Let $s > 1$ satisfy (1) and $t > 1$ satisfy for $\beta > 0$ the following condition

$$M_{t\varepsilon}/M_\varepsilon > 2(1+\beta). \tag{5}$$

Then

$$\exp\left(c(\beta)\frac{n}{t^2} \cdot \frac{M_\varepsilon^2}{\varepsilon^2}\right) \le N(K; \varepsilon D) \le \exp\left(C \cdot n s^2 \frac{M_\varepsilon^2}{\varepsilon^2}\right) \tag{6}$$

where C is a universal constant and $c(\beta) > 0$ depends on $\beta > 0$ only.

It is also easy to check that $c(\beta) \simeq c\beta^3$ (we have to use that dependence $\eta(\delta;c)$ on δ in Theorem C is $\approx$ const.δ^2).

3. We investigate next the "dual" quantity $N(D, \varepsilon K^\circ)$.

Theorem 3. a) Let $s \geq 1$ satisfy

$$(1^\circ) \qquad N(D, s\varepsilon K^\circ)^2 \leq N(D, \varepsilon K^\circ)$$

Then

$$N(D, \varepsilon K^\circ) \leq \exp(Cns^2(M_\varepsilon/\varepsilon)^2)$$

where C is a universal constant.

b) There exists a universal constant $a > 2$ such that

$$A_k(K)/M_\varepsilon \geq a$$

implies

$$2^k \leq N(D, \varepsilon K^\circ).$$

Proof. We follow the same line as in the proof of Theorem 1. To show the upper estimate a) we use Theorem B and the following submultiplicative inequality, valid for all $\varepsilon > 0$ and $\tau > 0$:

$$(7) \qquad N(D, \varepsilon K^\circ) \leq N(D, \operatorname{conv}(\frac{\varepsilon}{2} K^\circ \cup \frac{\tau}{2} D)) N(\tau D, \varepsilon K^\circ)$$

Indeed, setting $\tau = 1/s$, by (1°) and Theorem B we get

$$\begin{aligned} N(D, \varepsilon K^\circ) &\leq N(D, \operatorname{conv}(\frac{\varepsilon}{2} K^\circ \cup \frac{1}{2s} D))^2 = N(D, \frac{\varepsilon}{2} (K \cap s\varepsilon D)^\circ)^2 \\ &\leq \exp(Cn(M_{s\varepsilon}/\varepsilon)^2) \\ &\leq \exp(Cns^2(M_\varepsilon/\varepsilon)^2). \end{aligned}$$

To prove (7), set $N_1 = N(D, \frac{1}{2}(\varepsilon K^\circ + \tau D))$ and $N_2 = N(\tau D, \varepsilon K^\circ)$. There exist $x, \ldots, x_{N_1}$ in $\mathbb{R}^n$ and $z_1, \ldots, z_{N_2}$ in $\mathbb{R}^n$ such that $D \subset \bigcup_{j=1}^{N_1} (x_j + \frac{1}{2}(\varepsilon K^\circ + \tau D))$ and $\tau D \subset \bigcup_{i=1}^{N_2} (z_i + \varepsilon K^\circ)$. Combining these two coverings we get

$$D \subset \bigcup_{j=1}^{N_1} \bigcup_{i=1}^{N_2} \; (x_j + \frac{1}{2} z_i + \varepsilon K^\circ).$$

So $N(D, \varepsilon K^\circ) \le N_1 N_2 \le N(D, \mathrm{conv}\,(\frac{\varepsilon}{2}K^\circ \cup \frac{\tau}{2}D))\, N_2$, since $\mathrm{conv}(\frac{\varepsilon}{2}K^\circ \cup \frac{\tau}{2}D) \subset \frac{1}{2}(\varepsilon K^\circ + \tau D)$. This completes the proof.

Proof of b). Set $N = N(D, \varepsilon K^\circ)$. There exist $x_1, \dots, x_N$ in $\mathbb{R}^n$ such that $D \subset \bigcup_{i=1}^N (x_i + \varepsilon K^\circ)$. Then

$$D + \varepsilon K^\circ \subset \bigcup_{i=1}^{N} (x_i + 2\varepsilon K^\circ).$$

Note that $\varepsilon K_\varepsilon^\circ = \mathrm{conv}(D \cup \varepsilon K^\circ) \subset D + \varepsilon K^\circ$ and $K_\varepsilon^\circ \cap E = (P_E K_\varepsilon)^\circ$ for every subspace $E \in G_{n,k}$. Therefore

$$\varepsilon (P_E K_\varepsilon)^\circ \subset \bigcup_{1}^{N} (x_i + 2\varepsilon K^\circ) \cap E.$$

By Brunn Theorem, $\mathrm{Vol}_k[(x_i + 2\varepsilon K^\circ) \cap E] \le \mathrm{Vol}_k(2\varepsilon K^\circ \cap E)$. So $\varepsilon^k \mathrm{Vol}_k(P_E K_\varepsilon)^\circ \le N (2\varepsilon)^k \mathrm{Vol}_k(P_E K)^\circ$. Use now Santalo's inequality $\mathrm{Vol}_k(P_E K)^\circ \mathrm{Vol}_k(P_E K) \le (\mathrm{Vol}_k D_k)^2$. Then

$$\frac{\mathrm{Vol}_k P_E K}{\mathrm{Vol}_k D_k} \le 2^k \cdot N \frac{\mathrm{Vol}_k D_k}{\mathrm{Vol}_k (P_E K_\varepsilon)^\circ} \tag{8}$$

Our next step is the use of the inverse form of Santalo's inequality [B.M.], which states the existence of a universal number $c > 0$ such that

$$\mathrm{Vol}_k(P_E K_\varepsilon)^\circ \mathrm{Vol}_k(P_E K_\varepsilon) \ge c^k (\mathrm{Vol}_k D_k)^2.$$

Applying this inequality to (8) we have

$$\left(\frac{c}{2}\right)^k \frac{\mathrm{Vol}_k P_E K}{\mathrm{Vol}_k D_k} \le N \cdot \frac{\mathrm{Vol}_k P_E K_\varepsilon}{\mathrm{Vol}_k D_k} \tag{9}$$

for any $E \in G_{n,k}$. Integrating (9) over $G_{n,k}$ we receive

$$N \ge \left[\frac{c}{2} \frac{A_k(K)}{A_k(K_\varepsilon)}\right]^k \ge \left[\frac{c}{2} \frac{A_k(K)}{M_\varepsilon}\right]^k \tag{10}$$

(we use Alexandroff inequality on the last step). To finish the proof, take $a = 4/c$.

Remark 4. Beside the lower bound (10) we will note also another bound. From the covering $D \subset \bigcup_1^N (x_i + \varepsilon K^\circ)$, we have for every $E \in G_{n,k}$

$$D \cap E \subset \bigcup_{1}^{N} (x_i + \varepsilon K^\circ) \cap E$$

and, by Brunn Theorem

$$\mathrm{Vol}_k D_k \leq N \mathrm{Vol}_k(\varepsilon K^\circ \cap E) = \varepsilon^k \cdot N \cdot \mathrm{Vol}_k(K^\circ \cap E).$$

Using Santalo's inequality, we have

$$N \geq \frac{1}{\varepsilon^k} \frac{\mathrm{Vol}_k(P_E K)}{\mathrm{Vol}_k D_k}.$$

Integrating over $G_{n,k}$, we finally derive

$$N(D, \varepsilon K^\circ) \geq [A_k(K)/\varepsilon]^k.$$

□

In the same way, as we have obtained Corollary 2 from Theorem 1 and Remark 3, we obtain the next Corollary from Theorem 3 and Remark 3.

Corollary 4. Let $s > 1$ satisfy (1°) and $t > 1$ satisfy

$$M_{t\varepsilon}/M_\varepsilon > a$$

(where $a > 2$ is the universal constant from Theorem 3,b). Then

$$\exp\left(\frac{c_1}{t^2} \cdot n \frac{M_\varepsilon^2}{\varepsilon^2}\right) \leq N(D, \varepsilon K^\circ) \leq \exp\left(C_2 s^2 \cdot n \frac{M_\varepsilon^2}{\varepsilon^2}\right)$$

where $c_1 > 0$ and C_2 are universal constants.

References

[BZ] Y.D. Burago, V.A. Zalgaler. Geometric inequalities. "Nauka", Leningrad, 1980 (Russian).

[BM] J. Bourgain, V.D. Milman. New volume ratio properties for convex symmetric bodies in $\mathbb{R}^n$. Invent. Math., to appear.

[C] B. Carl. Entropy numbers, s-numbers and eigenvalue problem. J. of Funct. Anal. 41 (1980), pp. 290-306.

[Ch] G.D. Chakerian. Inequalities for the difference body of a convex body. Proc. Amer. Math. Soc. 18, No. 5 (1967), 879-884.

[MSch] V.D. Milman, G. Schechtman. Asymptotic theory of finite dimensional normed spaces. Springer-Verlag, Lecture Notes in Math., 1200, 156 pp. (1986).

[PT1] A. Pajor, N. Tomczak-Jaegermann. Subspaces of small codimension of finite-dimensional Banach spaces. Proc. AMS 97 (1986), 637-642.

[PT2] A. Pajor, N. Tomczak-Jaegermann. Remarques sur les nombres d'entropie d'un opérateur et de son transposé. C.R. Acad. Sci. Paris.

[S] L.A. Santaló. Integral geometry and geometric probability. Encyclopedia of Mathematics and its Applications, Addison-Wesley, 1976.

[Su] V.N. Sudakov. Gaussian random processes and measures of solid angles in Hilbert space. Soviet Math. Dokl. 12 (1971), 412-415.

[T] N. Tomczak-Jaegermann. Banach-Mazur distances and finite-dimensional operator ideals. To appear in Pitman.

AN APPLICATION OF INFINITE DIMENSIONAL HOLOMORPHY TO THE GEOMETRY OF BANACH SPACES

JONATHAN ARAZY
Department of Mathematics
University of Haifa
Haifa, 31999
ISRAEL

§1. Introduction

The purpose of these notes is to present the proof of the following result of W. Kaup and H. Upmeier (see also [BKU]):

1.1 Theorem [KU]: *Two complex Banach spaces whose open unit balls are biholomorphically equivalent are linearly isometric.*

The notes are written for non-specialists in infinite dimensional holomorphy who are interested in the geometry of Banach spaces. Along the way we survey some results concerning the Lie algebra $aut(D)$ of all complete holomorphic vector fields on the open unit ball D of the complex Banach space E, and its associated real Banach Lie group $Aut(D)$ consisting of all biholomorphic automorphisms of D. These results are due mainly to W. Kaup, H. Upmeier and J. P. Vigue, and are the extension to infinite dimensions of the works of E. Cartan and H. Cartan on bounded symmetric domains and of the biholomorphic automorphisms of bounded domains in the context of C^n. Naturally, this survey contains very few proofs. Then we give the full details of the fact that *the orbit of the origin under the group* $Aut(D)$ *is preserved under biholomorphic maps*. This is the main point in the proof of Theorem 1.1. From this the proof of Theorem 1.1 is reduced to showing that *biholomorphic maps which fix the origin are necessarily linear,* and this follows easily from the versions of Schwarz's lemma and Cartan's linearity theorem. One corollary of the theory is that *there exists a closed complex subspace* E_s *of* E *so that* $E_s \cap D = (Aut(D))(0)$. Thus $Aut(D)$ acts transitively on the open unit ball $E_s \cap D$ of E_s, and so $E_s \cap D$ is a bounded symmetric domain. The space E_s is called the *symmetric part of* E and its open unit ball $D_s = E_s \cap D$ is *the symmetric part of* D. In the last section we review some examples of spaces whose symmetric part is known, in particular the JB^**-triples* ($E_s = E$) and some spaces with an the *linear biholomorphic property* ($E_s = \{0\}$). We explain the role of the *contractive projection principle* in studying the symmetric parts of Banach spaces. Thullen's classification of Reinhardt domains in C^2 is discussed and the extension to infinite dimensions is described. It follows that the symmetric part of a Banach space with 1-unconditional basis is a band subspace which is naturally isometric to the c_0-sum of Hilbert spaces. It follows that rearrangement invariant Banach function spaces and unitary ideals have LBP provided they are not Hilbert spaces or C^*-algebras. Finally, we present an example of J.P. Vigue showing that E_s *need not be complemented in* E.

E. Cartan's classification of bounded symmetric domains in C^n is given in [CE]. For the works of H. Cartan on groups of biholomorphic automorphisms of domains in C^n see [CH1] and [CH2]. L. Harris (see [H1] and [H3]) extended the study of bounded symmetric domains and their connections to C^* algebras in the context of infinite dimensional spaces. He also extended Schwarz's lemma to the infinite dimensional setting, see [H2]. W. Kaup and H. Upmeier, and independently J. P. Vigue extended the works of E. Cartan, H. Cartan and L. Harris even further. The main point in their works is the connection between very general domains in complex Banach spaces and certain algebraic structure on the space, called the Jordan triple product. This triple product generalizes the binary product in C^*-algebras and Jordan algebras and is more suitable in many problems. For Kaup's works see [K1], [K2], [K3] and [K4]. The references to Upmeier's works are [U1] and [U2]. The book [U3] and the memoir [U4] survey the general perspective of the theory. Among Vigue's work we list first his thesis [V1] and the papers [V2], [V3], [V4] and [V5]. See also [IV].

The monograph [L] surveys the connection between bounded symmetric domains and Jordan triple systems in finite dimensional spaces. A detailed, elementary exposition of the theory of groups of biholomorphic automorphisms of bounded domains in Banach spaces and the connection to Jordan triple systems is found in [IS].

With the exception of Corollary 5.14, the results presented in the last section are not original. They are due besides Kaup, Upmeier and Vigue also to T.Barton, R. Braun, S. Dineen, Y. Friedman, B. Russo, L. L. Stacho and R. M. Timoney.

All these results deal with *complex* Banach spaces. H. Rosenthal (see [R1] and [R2]) studies the Lie algebra of linear skew Hermitian bounded operators on *real* Banach spaces.

The monographs [D1] and [FV] are general references to infinite dimensional holomorphy, and [LT1] and [LT2] to the geometry of Banach spaces.

Sections 2 and 4 below contains the details of the proof of Theorem 1.1. Section 3 contains the Lie-theoretic background. Section 5 can be considered as an appendix and it contains the information on the symmetric parts of certain families of Banach spaces.

We pass to notation and background material. In what follows E and F are complex Banach spaces. Let $n \geq 0$ and let

$$\bar{f} : \underbrace{E \times E \times \cdots \times E}_{n \text{ factors}} \to F$$

be a continuous, symmetric, multilinear map. The associated *homogeneous polynomial of degree n* is the restriction of $\bar{f}$ to the "diagonal", i.e. the map $f : E \to F$ defined by

$$f(z) = \bar{f}(\underbrace{z, z, \ldots, z}_{n \text{ terms}}), \quad z \in E.$$

For $n = 0$, $\bar{f}$ and f are interpreted as constant maps.

Let U be a domain in E (i.e. an open, connected subset). A function $f : U \to F$ is said to be *holomorphic* (or, *analytic)* if for each $z_0 \in U$ there is an open ball $B \subset U$ with center z_0 and a sequence $\{f_n\}_{n=0}^{\infty}$, where f_n is a homogeneous polynomial of degree n, so that

$$f(z) = \sum_{n=0}^{\infty} f_n(z - z_0), \quad z \in B. \tag{1.1}$$

The series (1.1) is called the *Taylor series* of f at z_0. It is known that f is holomorphic in U if and only if the *Frechet derivative* of f at z_0, denoted by $f'(z_0)$, exists for every $z_0 \in U$. One observes that if f is given by (1.1) then

$$f'_n(w)z = n\,\bar{f}_n(w, w, \ldots, w, z).$$

In particular,

$$f'_n(z)(z) = n\, f_n(z)$$

and

$$f'(z_0)z = f_1(z).$$

If $f : U \to F$ is holomorphic, one-to-one, $f(U)$ is open in F and $f^{-1} : f(U) \to U$ is holomorphic, then f is said to be *biholomorphic,* and the sets U and $f(U)$ are *biholomorphically equivalent.* It is known that f is biholomorphic if and only if $f'(z_0)$ is an isomorphism of E onto F for every $z_0 \in U$.

We remark that, unlike the finite dimensional case, it is an open problem whether a one-to-one holomorphic function from U onto an open subset $f(U)$ of F is biholomorphic.

A ball $B \subset U$ is *completely interior in* U *if* $dist(B, \partial U) > 0$. *If* $f : U \to F$ *is* holomorphic we set

$$\|f\|_B = \sup_{z \in B} \|f(z)\|$$

Let $H(U,F)$ denote the space of all holomorphic functions from U into F, and let $H_0(U,F)$ be the subspace consisting of all $f \in H(U,F)$ which are bounded on each ball B completely interior to U. $H_0(U,F)$ is endowed with the *topology of local uniform convergence* defined by the family of seminorms $\|\cdot\|_B$, where B ranges over all balls completely interior to U.

It is known that this topology is metrizable and complete on $H_0(U,F)$.

Remark: Let B_1, B_2 be two balls completely interior in U, and let $M \in (0,\infty)$. Then the topologies induced by the norms $\|\cdot\|_{B_1}$, $\|\cdot\|_{B_2}$ are equivalent on each set

$$H_M(U,F) = \{f \in H_0(U,F) ; \sup_{z \in U} \|f(z)\| \leq M\}.$$

Indeed, in case of concentric balls this follows from the Hadamard's three circles theorem. The general case follows from the observation that there exist open balls $C_1, \ldots, C_n$ with $C_1 = B_1$ and $C_n = B_2$, so that for some $\varepsilon > 0 : (C_j)_\varepsilon$ are completely interior in U and $(C_j)_\varepsilon \supset C_{j+1}$, $j = 1,2, \ldots, n-1$. Here A_ε denotes the ε-neighborhood of a set A.

§2 The proof of Theorem 1.1 in case $f(0) = 0$

Let E,F be complex Banach spaces with open unit balls $D(E)$ and $D(F)$. If $f : D(E) \to D(F)$ is biholomorphic (onto) then for each $z \in D(E)$, $f'(z)$ is a linear isomorphism of E onto F with inverse $(f^{-1})'(f(z))$. But in general $f'(z)$ need not be an isometry, as the following example shows. Let Δ be the open unit disc in the complex plane C, let $\varphi_a : \Delta \to \Delta$ be defined by $\varphi_a(z) = (a-z)(1-\bar{a}z)^{-1}$. Then $\varphi_a \in Aut(\Delta)$ and $\varphi'_a(0) =- 1+|a|^2$. So $\varphi'_a(0) : C \to C$ is not an isometry if $a \neq 0$. It is well known that if $\varphi \in Aut(\Delta)$ and $\varphi(0) = 0$ then φ is a rotation. The following lemma show that this is the case in general.

2.1 Lemma: *Let $f : D(E) \to D(E)$ be biholomorphic, onto. If $f(0) = 0$ then f is the restriction to $D(E)$ of a linear isometry of E onto F.*

Proof: Fix $z \in E$, and $x^* \in F^*$ with $\|z\| = 1 = \|x^*\|$, and consider $\varphi : \Delta \to \Delta$ defined via

$$\varphi(\lambda) = x^*(f(\lambda z)).$$

By the Schwarz lemma $|\varphi'(0)| \leq 1$. This holds for all x^* and z, and so

$$\|f'(0)\| \leq 1.$$

Similarly

$$\|(f^{-1})'(0)\| \leq 1$$

thus $f'(0)$ is a linear isometry of E onto F. (Note that this already proves Theorem 1.1 in the case $f(0) = 0$). Let

$$\psi = (f^{-1})'(0) \circ f : D(E) \to D(E)$$

Then ψ is biholomorphic and

$$\psi(0) = 0, \quad \psi'(0) = I_E.$$

The following theorem yields $\psi = I_{E_{|D}}$ and hence $f = f'(0)$, thus completely the proof of the Lemma.

2.2 Theorem: *(H. Cartan's Linearity Theorem)*

Let D be a bounded domain in a complex Banach space E, let $\psi : D \to D$ be holomorphic and assume that for some $z_0 \in D$.

$$\psi(z_0) = z_0, \quad \psi'(z_0) = I_E.$$

Then $\psi = I_{E_{|D}}$.

We sketch the proof. Without loss of generality $z_0 = 0$. If $\psi \neq I_E$ then its Taylor series about 0 is

$$\psi(z) = z + \psi_m(z) + \cdots \text{(higher order terms)} \cdots$$

where $m \geq 2$ and $0 \neq \psi_m$ is a homogeneous polynomial of degree m. The n'th iterate $\psi^{[n]}$ has Taylor series

$$\psi^{[n]}(z) = z + n\,\psi_m(z) + \cdots \text{(higher order terms)} \cdots$$

But for z near 0

$$n\psi_m(z) = \frac{1}{2\pi} \int_0^{2\pi} \psi^{[n]}(e^{i\theta} z) e^{-im\theta} d\theta.$$

Hence

$$\| n\psi_m(z) \| \leq \sup\{ \| w \|, \ w \in D \} < \infty.$$

Letting $n \to \infty$ we see that $\psi_m(z) = 0$ in a neighborhood of 0, and so $\psi_m(z) = 0$ identically. This contradiction completes the proof.

We denote by $G(E) = Aut(D(E))$ resp. $G(F) = Aut(D(F))$ the group of all biholomorphic automorphisms of $D(E)$, respectively $D(F)$. Lemma 2.1 reduces Theorem 1.1 to the following result.

2.3 Theorem: *Let $f : D(E) \to D(F)$ be biholomorphic and onto. Then*

$$f(G(E)(0)) = G(F)(0).$$

Indeed, if this holds then there exists $\varphi \in G(F)$ so that $f(0) = \varphi(0)$. Thus $\psi = \varphi^{-1} \circ f : D(E) \to D(F)$ is biholomorphic and $\psi(0) = 0$. By Lemma 2.1 $\psi = \psi'(0)_{|D(E)}$ and $\psi'(0)$ is a linear isometry of E onto F. This completes the proof of Theorem 1.1 modulo Theorem 2.3.

The proof of Theorem 2.3 is based on a *biholomorphic characterization of the orbit* $G(E)(0)$, namely on establishing a property of $G(E)(0)$ which is preserved by biholomorphic maps. This is done in §4, while §3 is devoted to some Lie theoretic background.

§3. The Lie-theoretic background

Let D be the open unit ball of the complex Banach space E. Every holomorphic function $h : D \to E$ induces a *holomorphic vector field* X, written symbolically as $X = h(z)\frac{\partial}{\partial z}$, which is a differential operator on the space $H(D,E)$ of all holomorphic functions $h : D \to E$, defined via

$$(Xf)(z) = (h(z)\frac{\partial}{\partial z})f(z) = f'(z)(h(z)),\ z \in D.$$

The *flow of* X *through the point* $z_0 \in D$ is the unique solution $\varphi = \varphi_X : J_{z_0} \to D$ of the initial value problem

$$\begin{cases} \frac{d}{dt}\varphi(t) = h(\varphi(t)), & t \in J_{z_0} \\ \varphi(0) = z_0 \end{cases}.$$

Here J_{z_0} is the maximal open interval containing $t_0 = 0$ in which the solution exists. The vector field X is said to be *complete* if $J_{z_0} = \Re$ for every $z_0 \in D$. In this case one gets a function

$$\varphi = \varphi_X : \Re \times D \to D$$

which is holomorphic in each variable, and satisfies for all $z \in D$

$$\begin{cases} \frac{\partial}{\partial t}\varphi(t,z) = h(\varphi(t,z)), & t \in \Re \\ \varphi(0,z) = z \end{cases}.$$

Also, the function $\varphi_t : D \to D$, defined by $\varphi_t(z) = \varphi(t,z)$, belongs to $Aut(D)$ and

$$\varphi_{t+s} = \varphi_t \circ \varphi_s.$$

Thus $\{\varphi_t\}_{t\in\Re}$ is a one-parameter subgroup of $Aut(D)$ whose *generator* is X, that is for every $f \in H(D,E)$

$$X(f)(z) = \frac{\partial}{\partial t} f(\varphi_t(z))|_{t=0}.$$

One denotes

$$\exp(tX)(z) = \varphi_t(z) = \varphi(t,z).$$

In particular $\exp(X) = \varphi_1$ is the *exponential* of X.

3.1 Definition: The set of all complete holomorphic vector fields on D is denoted by $aut(D)$.

The *commutator* of $X, Y \in Aut(D)$ is defined by

$$([X,Y]f)(z) = Y(Xf)(z) - X(Yf)(z).$$

If $X = h(z)\frac{\partial}{\partial z}$, $Y = k(z)\frac{\partial}{\partial z}$ then

$$[X,Y] = (h'(z)k(z) - k'(z)h(z))\frac{\partial}{\partial z}.$$

3.2 Theorem: *Let B be any open ball completely interior to D. Then $aut(D)$ is a real Banach Lie-algebra with respect to the commutator product and the norm $\|\cdot\|_B$.*

Let us just indicate why $Aut(D)$ is closed under the operations of addition and commutation. One first identifies $aut(D)$ with the tangent space of $Aut(D)$ at the identity element $e = id_D$, with respect to the topology of local uniform convergence:

$$aut(D) = T_e(Aut(D)).$$

The meaning of this is that if $t \to g_t$, $|t| < \varepsilon$, is a smooth curve in $Aut(D)$, $g_0 = e$, and

$$h(z) = \frac{\partial}{\partial t} g_t(z)_{|t=0}$$

then $X = h(z)\frac{\partial}{\partial z} \in aut(D)$.

Next, if $X = h(z)\frac{\partial}{\partial z}$, $Y = k(z)\frac{\partial}{\partial z}$ are in $aut(D)$, then it is elementary to prove that

$$\frac{\partial}{\partial t} \varphi_X(t, \varphi_Y(t,z))_{|t=0} = h(z) + k(z).$$

Hence $X + Y \in aut(D)$. To show that $[X,Y] \in aut(D)$ we consider

$$g_t(z) = \varphi_X(t, \varphi_Y(t, \varphi_X(-t, \varphi_Y(-t,z)))).$$

Clearly $g_0(z) = z$, $\frac{\partial}{\partial t} g_t(z)_{|t=0} = 0$. Moreover,

$$\lim_{t \to 0} \frac{g_t(z) - z}{t^2} = h'(z)k(z) - k'(z)h(z),$$

showing that $[X,Y] \in aut(D)$.

The Lie algebra $aut(D)$ is *purely real* in the sense that

$$aut(D) \cap i\, aut(D) = \{0\}.$$

Indeed, let $X = h(z)\frac{\partial}{\partial z} \in (aut(D) \cap i\, aut(D))$. Since $[X, iX] = 0$ the one-parameter groups $\{\exp(tX)\}_{t \in \mathfrak{R}}$ and $\{\exp s(iX)\}_{s \in \mathfrak{R}}$ commute. Thus for fixed $z \in D$

$$F(t+is) = \exp(tX)(\exp s(iX)(z))$$

is a bounded entire function in $t + is$. By Liouville's theorem, F is a constant. Thus $0 = F'(0) = \frac{\partial}{\partial t} F_{|t=s=0} = h(z)$. Since $z \in D$ is arbitrary, $X = 0$ as asserted.

The following is classical.

3.3 Theorem (H.Cartan's Uniqueness Theorem): *Fix $a \in D$.*

(i) Suppose that $\varphi, \psi \in Aut(D)$ satisfy

$$\varphi(a) = \psi(a) \quad , \varphi'(a) = \psi'(a)$$

then $\varphi = \psi$

(ii) Suppose $h\frac{\partial}{\partial z}$, $k\frac{\partial}{\partial z} \in aut(D)$ satisfy

$$h(a) = k(a), \quad h'(a) = k'(a)$$

then $h = k$.

Proof: (i) Let $\sigma = \varphi \circ \psi^{-1}$. Then $\sigma(a) = a$ and $\sigma'(a) = I_E$. Hence $\sigma = I_{E_{|D}}$ by Theorem 2.2.

(ii) Without loss of generality assume $k = 0$. Let $X = h\frac{\partial}{\partial z}$, let $\varphi = \varphi_X : \mathfrak{R} \times D \to D$ be the flow of X, and let $\varphi_t(z) = \varphi(t,z)$. Clearly, $\varphi_t(a) = a$ for every $t \in \mathfrak{R}$ because this function solves the initial value problem

$$\begin{cases} \frac{\partial}{\partial t}\varphi_t(a) = h(\varphi_t(a)) \\ \varphi_0(a) = a \end{cases}.$$

Next, we claim that $\frac{\partial}{\partial z}\varphi_t(z)_{|z=a} = I_E$ for every $t \in \mathfrak{R}$. Having this we get from the first part that $\varphi_t(z) = z$ for all t and z, and so $h(z) = 0$ identically. To prove the claim we consider $\frac{\partial^2}{\partial t \partial z}\varphi_t(z)$ at $z = a$. First

$$\frac{\partial^2 \varphi_t(z)}{\partial t \partial z} = \frac{\partial}{\partial z}(\frac{\partial}{\partial t}\varphi_t(z))$$

$$= \frac{\partial}{\partial z} h(\varphi_t(z))$$

$$= h'(\varphi_t(z)) \frac{\partial}{\partial z} \varphi_t(z)$$

and since $h'(a) = 0$, we get

$$\frac{\partial^2}{\partial t \partial z} \varphi_t(z)_{|z=a} = 0.$$

Hence

$$\frac{\partial}{\partial t}(\frac{\partial}{\partial z} \varphi_t(z)_{|z=a}) = 0.$$

So $\frac{\partial}{\partial z}\varphi_t(z)_{|z=a}$ is constant with respect to t ,hence

$$\frac{\partial}{\partial z}\varphi_t(z)_{|z=a} = \frac{\partial}{\partial z} \varphi_0(z)_{|z=a}$$

$$= \frac{\partial}{\partial z} z_{|z=a} = I_E.$$

This establishes the claim and completes the proof.

□

The following fundamental result is due to J.P. Vigue. It is the topological version of Cartan's Uniqueness Theorem. Here $L(E)$ denotes the space of all bounded linear operators on E.

3.4 Theorem [V1]: *Fix $a \in D$. Then the map $T : aut(D) \to E \oplus L(E)$ defined via*

$$T(h\frac{\partial}{\partial z}) = (h(a), h'(a))$$

is an isomorphism into. In particular, the image of T is closed in $E \oplus L(E)$. Thus the norms $\|h\|_B$ and $\|h(a)\|_E + \|h'(a)\|_{L(E)}$ are equivalent on $aut(D)$.

The following simple result is the key for the analysis of the structure of $aut(D)$ and is the basis for the connection between the holomorphic structure of D and the structure of E.

3.5 Lemma: *Let $h\frac{\partial}{\partial z} \in aut(D)$.*

(i) h is quadratic in the sense that its Taylor series at 0 is $h = h_0 + h_1 + h_2$ with $\{h_n\}$ homogeneous polynomial of degree n;

(ii) $h_1\frac{\partial}{\partial z}$ and $(h_0+h_2)\frac{\partial}{\partial z}$ belong to $aut(D)$

(iii) $h_0 = 0$ *if and only if* $h_2 = 0$.

Proof: D is circular about 0 so the rotations

$$\varphi_t(z) = e^{it} z, \quad t \in \Re, \ z \in D$$

form a continuous one-parameter subgroup of G. The infinitesimal generator is the vector field $u \frac{\partial}{\partial z} \in aut(D)$ with $u(z) = iz$. Let $T : aut(D) \to aut(D)$ be defined via

$$Th = [u, h]$$

where we drop the "$\frac{\partial}{\partial z}$" notation for convenience. Clearly, T is continuous with respect to local uniform convergence. Moreover, if h_n is a n-homogeneous polynomial then

$$(Th_n)(z) = u'(z)h_n(z) - h'_n(z)u(z) = i(1-n)h_n(z).$$

Let $h(z) = \sum_{n=0}^{\infty} h_n(z)$, then

$$(Th)(z) = \sum_{n=0}^{\infty} i(1-n)h_n(z).$$

If $p(x)$ is a polynomial with real coefficients then $p(T) : aut(D) \to aut(D)$ and

$$p(T)h = p(T)(\sum_{n=0}^{\infty} h_n) = \sum_{n=0}^{\infty} p(i(1-n))\, h_n.$$

Taking $p(x) = x(1+x^2)$ we get that

$$p(T)h = i \sum_{n=0}^{\infty} n(n-1)(n-2)h_n.$$

The vector field $p(T)h$ belongs to $aut(D)$, it vanishes at 0 and has vanishing derivative there. By Theorem 3.3 (ii), $p(T)h = 0$. so $h_n = 0$ for $n \geq 3$. Next, take $p(x) = 1+x^2$. Then the same argument yields

$$p(T)h = \sum_{n=0}^{2} n(n-2)h_n = h_1 \in aut(D).$$

Also, $h_0 + h_2 = h - h_1 \in aut(D)$. Finally, if $h_0 = 0$ then $h_2 = h - h_1 \in aut(D)$ and $h_2(0) = 0$, $h'_2(0) = 0$. Hence $h_2 = 0$ by Theorem 3.3(ii). And if $h_2 = 0$ then $h_0 \in aut(D)$. But the integral curves of a constant vector field are line segments, so h_0 is incomplete unless it is zero. □

3.6 Theorem [KU],[BKU]: *Let*

$$E_s = \{h(0);\ h\frac{\partial}{\partial z} \in aut(D)\}.$$

Then E_s *is a closed, complex linear subspace of* E.

Proof: The fact that E_s is a closed, real linear subspace of E follows easily from Theorem 3.4. To prove that it is complex linear we apply the technique of the proof of Lemma 3.5. Let $h = h_0 + h_1 + h_2 \in aut(D)$, then

$$Th = [u, h] = i(h_0 - h_1) \in aut(D).$$

So $h_0 \in E_s$ implies $ih(0) \in E_s$ and so E_s is complex linear.

□

3.7 Definition: Let $a \in E_s$. We denote by $h_a(z) = a - q_a(z)$ the unique element of $aut(D)$ satisfying $h_a(0) = a$ and $h'_a(0) = 0$. Here q_a is a homogeneous polynomial of degree 2. We denote the associated symmetric bilinear map by $q_a(z,w)$. The *partial triple product* on E is the map

$$\{\ \} : E \times E_s \times E \to E$$

defined via

$$\{z\ a\ w\} = q_a(z,w).$$

Remark: The map $a \to q_a$ is clearly real-linear. The proof of Theorem 3.6 yields $q_{ia} = -iq_a$. Hence $a \to q_a$ is a conjugate-linear map. Thus the partial triple product $\{z\ a\ w\}$ is bilinear and symmetric in z and w and conjugate-linear in a. See [K1], [K3] and [V2] for further investigation.

The *exponential map* $\exp : aut(D) \to Aut(D)$ is not injective. But it becomes so when restricted to an appropriate neighborhood of 0, in which it is given by a convergent power series (and so the inverse map is given by the inverse power series), see [IS, chapter 6]. Precisely

3.8 Lemma: *Let B be an open ball in D and let $\delta > 0$ be so that $B_\delta = \{z \in D\ ;\ dist(z,B) < \delta\}$ is completely interior in D. Let*

$$\mathcal{M} = \{h\frac{\partial}{\partial z} \in aut(D)\ ;\ \|h\|_{B_\delta} < \frac{\delta}{2e}\}$$

(i) for any $X = h\frac{\partial}{\partial z} \in \mathcal{M}$ the series

$$\sum_{n=0}^{\infty} \frac{1}{n!} X^n(id_D)$$

converges to $\exp X$ *in the norm* $\|\cdot\|_B$

(ii) let $\mathcal{N} = \exp(\mathcal{M})$. Then the map $\exp : \mathcal{M} \to \mathcal{N}$ *is one-to- one (hence a bijection) and is real analytic when $\mathcal{N}$ is taken with the norm $\|\cdot\|_B$.*

3.9 Definition: *The analytic topology on* $Aut(D)$ is defined in terms of the following system of open neighborhoods of the identity

$$\{\exp(\frac{1}{n}\ \mathcal{M};\ n = 1,2,\cdots\}$$

where $\mathcal{M}$ is the set defined in Lemma 3.8.

Remark: The analytic topology is finer than (and in general different from) the topology of local uniform convergence.

3.10 Definition: A Hausdorff topological space M is an *analytic Banach manifold* over the field K (of either the real or complex scalars) if

(i) for every $x \in M$ there exists an open neighborhood U in M and a homeomorphism h_U of U onto some open subset W_U in a Banach space over K,

(ii) If U_1, U_2 are open subsets of M which arise in this way and $U_1 \cap U_2 \neq \phi$, then the map $h_{U_2} \circ h_{U_1}^{-1} : W_1 \to W_2$ is bianalytic.
(If $K = \mathfrak{R}$ than "analytic" means "real analytic").

3.11 Definition: A *submanifold* N of an analytic Banach manifold M over the field K is a subset having the following property:

For every $x \in N$ there exists an open neighborhood U of x in M and an open neighborhood W of 0 in some Banach space X over K, a K-linear closed subspace Y of X and a bianalytic map h of U onto W such that

$$h(U \cap N) = W \cap Y.$$

We remark that the neighborhood U of x is taken in M rather than N. Thus the topology of N is the relative topology induced from M (in some books such submanifolds are called "imbedded" submanifolds).

3.11 Definition: A *Banach Lie-group* over K is a topological group G which is also a Banach manifold over K, such that the algebraic operations (product, inverse) are analytic (from $G \times G$ into G and from G into G respectively).

The following fundamental result is due to H. Upmeier [U1], [U2] and J.P. Vigue [V1] independently. See also [IS, chapter 6].

3.12 Theorem: *Let D be the open unit ball of the complex Banach space E. Then* $G = Aut(D)$, *endowed with the analytic topology, is a real Banach Lie- group.*

Let G^0 denote the connected component of the identity in $G = Aut(D)$. Clearly, G^0 is a closed, normal subgroup of G. Let $E_s = aut(D)(0)$ be the subspace of E considered in Theorem 3.6.

3.13 Theorem: (a) *The set* $D_s = E_s \cap D$ *is invariant under every member of* G.

(b) $G^0(0)$ *is an open neighborhood of* 0 *in* D_s.

The proof is based on differentiation of the evaluation map

$$G^0 \ni \psi \to \psi(0) \in D_s$$

and the application of Theorem 3.4 and the implicit function theorem.

§4. Biholomorphic characterization of the orbit of the origin under the group of biholomorphic automorphisms.

Let E be a complex Banach space, and let D be its open unit ball. We let $G = Aut(D)$ be the group of all biholomorphic automorphisms of D endowed with the analytic topology, and let G^0 be the connected component of the identity in G. $E_s = aut(D)(0)$ and $D_s = E_s \cap D$.

4.1 Definition:

$\Omega = \Omega(E) = \{z \in D \; ; G^0(z)$ is a closed complex submanifold of $D\}$.

4.2 Main Lemma: $G(0) = G^0(0) = \Omega = D_s$.

This lemma provides a biholomorphic characterization of the orbit of the origin $G(0)$ and thus proves Theorem 2.3, hence Theorem 1.1.

Indeed, assuming Lemma 4.2, let $f : D(E) \to D(F)$ be biholomorphic. Denote $G(E) = Aut(D(E))$, and $G(F) = Aut(D(F))$. Then the map $\varphi \to f \circ \varphi \circ f^{-1}$ is an isomorphism of $G(E)$ onto $G(F)$ as real Banach Lie groups. In particular

$$f \circ G^0(E) \circ f^{-1} = G^0(F).$$

If $z \in G(E)(0) = \Omega(E)$, then $G^0(E)(z)$ is a closed, complex submanifold of $D(E)$. Hence $f(G^{(0)}(E)(z))$ is a closed, complex submanifold of $D(F)$. But

$$f(G^0(E)(z)) = (f \circ G^0(E) \circ f^{-1})(f(z))$$
$$= G^0(F)(f(z)).$$

So $f(z) \in \Omega(F) = G(F)(0)$. It follows that $f(G(E)(0) \subseteq G(F)(0)$, and applying the same argument with f^{-1} we get $f(G(E)(0)) = G(F)(0)$ as desired.

Proof of the Main Lemma 4.2.

Step 1: $\Omega \subseteq G^0(0)$.

It is enough to show that

$$z \in \Omega \setminus \{0\} \quad \text{implies} \quad 0 \in G^0(z)$$

because $0 \in G^0(z)$ if and only if $z \in G^0(0)$. Let $[z]$ be the one-dimensional complex subspace generated by z. $G^0(z)$ is closed in D, so $[z] \cap G^0(z)$ is closed in $[z] \cap D$. $[z] \cap G^0(z)$ contains $e^{i\theta}z$ for all $\theta \in [0,2\pi)$, since G^0 contains the rotation subgroup.

Claim: $[z] \cap G^0(z)$ is open in $[z] \cap D$.

Having this we clearly get $[z] \cap G^0(z) = [z] \cap D$, since $[z] \cap D$ is connected. In particular, $0 \in G^0(z)$ as desired.

To prove the claim apply thc fact that $G^0(z)$ is a complex submanifold. Let U be an open ball contained in D with center z, W an open neighborhood of 0 in E,Y a closed complex linear subspace of E and let $h : U \to W$ be a biholomorphic map satisfying $h(U \cap G^0(z)) = W \cap Y$. The open subset $[z] \cap U$ of $[z]$ contains $\Gamma_\varepsilon = \{e^{i\theta}z;\ |\theta| < \varepsilon\}$ for some $\varepsilon > 0$, and the holomorphic function

$$h_1 = h_{|[z] \cap U}$$

maps Γ_ε into the subspace Y. Let $\pi : E \to E/Y$ be the canonical quotient map, then $h_2 = \pi \circ h_1$ is holomorphic in $[z] \cap U$ and $h_{2|\Gamma_\varepsilon} \equiv 0$. It follows by ordinary function theory that $h_2 \equiv 0$, and so

$$h_1([z] \cap U) = h([z] \cap U) \subseteq W \cap Y.$$

Applying h^{-1} we get

$$[z] \cap U \subseteq G^0(z) \cap U.$$

so $[z] \cap U \subseteq [z] \cap G^0(z)$, and hence z is an interior point of $[z] \cap G^0(z)$. If

$w \in [z] \cap G^0(z)$, $w \neq 0$, then w is an interior point of $[w] \cap G^0(w) = [z] \cap G^0(z)$. Thus $[z] \cap G^0(z)$ is an open subset of $[z] \cap D$. This proves the claim and thus completes the proof of step 1.

Step 2: $G^0(0) = D_s$.

By Theorem 3.13 $G^0(0)$ is an open subset of D_s. Also, D_s is invariant under all members of G^0. Therefore G^0 is identified naturally with a subgroup of $Aut(D_s)$. To prove that $G^0(0)$ is all of D_s, it is enough to show that $G^0(0)$ is closed in D_s, and then apply the connectivity of D_s. To this end we need

4.3 Lemma: *Let Z be a complex Banach space with unit ball U. Then there exists a $Aut(U)$-invariant metric ρ on U, such that the norm topology and the topology induced by ρ coincide on U.*

Assuming this we continue the proof with $Z = E_s$, $U = D_s$. Since $G^0(0)$ is open in D_s, there exists $\varepsilon > 0$ so that the ρ-ball

$$B_\rho(0,\varepsilon) = \{x \in D_s \; ; \rho(0,x) < \varepsilon\}$$

is contained in $G^0(0)$. But ρ is G^0-invariant, and so

$$B_\rho(z,\varepsilon) = \{x \in D_s \; ; \rho(z,x) < \varepsilon\} \subseteq G^0(0)$$

for every $z \in G^0(0)$. This clearly implies that $G^0(0)$ is closed in D_s, hence $D_s = G^0(0)$.

Proof of Lemma 4.3: Let ρ_0 be the Poincare metric in Δ, namely

$$\rho_0(a,b) = \frac{1}{2}\log\left(\frac{1+\left|\dfrac{a-b}{1-\bar{a}b}\right|}{1-\left|\dfrac{a-b}{1-\bar{a}b}\right|}\right) \; ; \qquad a,b \in \Delta$$

ρ_0 is the unique (up to a constant multiple) $Aut(\Delta)$- invariant metric in Δ. We take ρ to be the *Caratheordory metric* on U, that is

$$\rho(z,w) = \sup\{\rho_0(f(z),f(w)); f \in H(U,\Delta)\}.$$

It is clear that ρ is $Aut(U)$-invariant, symmetric and satisfies the triangle inequality. The fact that ρ is locally equivalent to the distance given by the norm (and hence ρ is also a metric, not just a semi metric) is due to the fact that ρ_0 is locally equivalent to the Euclidean metric on Δ. Indeed, $\rho_0(a,b) \geq |a-b|$, and so

$$p(z,w) \geq \sup_{z^* \in Z^*, \|z^*\|=1} \rho_0(z^*(z), z^*(w))$$

$$\geq \sup_{z^* \in Z^*, \|z^*\|=1} |z^*(z) - z^*(w)| = \|z-w\|.$$

Conversely, let $z \in U$, and let $0 < \delta$ be so that $B = \{w \in Z \,;\, \|z-w\| < \delta\} \subseteq U$. Let $w \in B \setminus \{z\}$, then the intersection of B with the hyperplane through z and w is identified with $\delta\Delta$, and so

$$\rho(z,w) \leq \rho_0(\frac{\|z-w\|}{\delta}, 0) \leq C_\delta \|z-w\|.$$

□

Step 3: $G(0) \subseteq \Omega$

We begin with the remark that the normality of G^0 in G implies that Ω is G-invariant. This was proved in the discussion following the statement of Lemma 4.2. By Step 2 and Theorem 3.13, $G^0(0) = D_s$ is a closed complex submanifold of D. Hence $0 \in \Omega$. Since Ω is G-invariant we get $G(0) \subseteq \Omega$, completing the proof.

Step 4: Conclusion of the proof of Lemma 4.2.

We have,

$$
\begin{aligned}
G(0) &\subseteq \Omega && \text{(Step3)} \\
&\subseteq G^0(0) && \text{(Step1)} \\
&= D_s && \text{(Step2)} \\
&\subseteq G(0) && (G^0 \subseteq G).
\end{aligned}
$$

Therefore equality holds everywhere and the proof is complete.

□

4.4 Corollary : *Let* $G_0 = \{\varphi \in G \,;\, \varphi(0) = 0\}$ *be the isotropy subgroup at* 0. *Then*

$$G = G^0 G_0 = G_0 G^0.$$

Proof: Let $\psi \in G$. By Lemma 4.2 there exists $\psi_1 \in G^0$ so that $\psi(0) = \psi_1(0)$. Therefore $\psi_2 = \psi_1^{-1}\psi \in G_0$ and so $\psi = \psi_1\psi_2$. This shows that $G = G^0 G_0$. The normality of G^0 in G yields now that $G = G_0 G^0$.

§5. The symmetric part of a Banach space

Let E be a complex Banach space with an open unit ball D. Let $G = Aut(D)$ be the group of all biholomorphic automorphisms of D and let G^0 the connected component of the identity in G. $\underline{g} = aut(D)$ denotes the Lie algebra of all complete, holomorphic vector fields on D.

5.1 Definition: The *symmetric part of D* is

$$D_s = G(0) = G^0(0).$$

The *symmetric part of E* is the closed, complex subspace of E

$$E_s = \underline{g}(0) = \{h(0) \,;\, h\frac{\partial}{\partial z} \in \underline{g}\}.$$

Notice that according to Lemma 4.2

$$D_s = E_s \cap D.$$

So D_s is the open unit ball of E_s. D_s is invariant under both G and G^0, thus they are identified with subgroups of $Aut(D_s)$. D_s is *homogeneous,* i.e. $Aut(D_s)$ acts on it *transitively,* because G and G^0 act transitively. Moreover D_s is symmetric in the sense that for each $z \in D_s$ there exists a $s_z \in Aut(D_s)$ so that

$$s_z(z) = z, \; s_z^2 = \text{identity, and } \; s'_z(z) = -I_E.$$

s_z is called the *symmetry at z*; according to Theorem 3.3 (Cartan's Uniqueness Theorem) s_z is the unique member of $Aut(D_s)$ with these properties. Indeed $s_0 = -id_{D_s}$ is the symmetry at 0. If $\varphi \in Aut(D_s)$ and $z = \varphi(0)$, then $s_z = \varphi \circ s_0 \circ \varphi^{-1}$.

The following fundamental result is due to L. L. Stacho [S2].

5.2 Theorem: *(contractive projection principle) Let E be a complex Banach space and let $P : E \to E$ a contractive (i.e. norm 1) projection. Let $h\frac{\partial}{\partial z}$ be a complete holomorphic vector field on D and define*

$$\tilde{h} = Ph_{|P(E)\cap D} : P(E) \cap D \to P(E).$$

Then $\tilde{h}\frac{\partial}{\partial z}$ is a complete, holomorphic vector field on $P(E) \cap D$.

Theorem 5.2 is a corollary of a very general result proved in [S2]. In our case a more direct proof can be given based on the following characterization of complete, holomorphic vector fields:

Let $X = h\frac{\partial}{\partial z}$ *be a holomorphic vector field on* D. *Then* X *is complete if and only if* h *extends holomorphically to a neighborhood of* $\bar{D}$ *and* X *is tangent to* ∂D *in the sense that whenever* $x \in E$, $x^* \in E^*$ *satisfy* $\|x\| = 1 = \|x^*\|$ *and* $x^*(x) = 1$, *then* $x^*(h(x)) = 0$. [See also [U4, Lecture 4].

5.3 Corollary: *Let* P *be a contractive projection in* E. *Then*

$$P(E_s) \subseteq (P(E))_s.$$

Proof: Let $h\frac{\partial}{\partial z} \in aut(D)$, and let $\tilde{h} = Ph_{|P(E) \cap D}$. Then by Theorem 5.2 $\tilde{h}\frac{\partial}{\partial z} \in aut(P(E) \cap D)$ and

$$\tilde{h}(0) = Ph(0) \in (P(E))_s.$$

Since $h(0)$ is an arbitrary point in E_s, we get $P(E_s) \subset (P(E))_s$.

□

We turn to some examples of Banach spaces whose symmetric parts are known.

JB^* - TRIPLE SYSTEMS

Our first example, in fact a family of examples, deals with the case where $E_s = E$ and $D_s = D$, that is E and D are symmetric. In this case the partial triple product introduced in §3 via $\{xyz\} = q_y(x,z)$ becomes a triple product, namely it is defined now for all $x,y,z \in E$. One denotes for $x,y \in E$ $(x \,\Box\, y)z = \{xyz\}$. In his fundamental work [K1] (see also [V2]) W. Kaup establishes the following properties of the map $\Box : E \times E \to L(E)$.

(i) $\Box$ *is continuous and sesquilinear;*

(ii) The Jordan Triple Identity: for every $x,y,u,v \in E$,

$$[x \,\Box\, y, u \,\Box\, v] = \{xyu\} \,\Box\, v - u \,\Box\, \{vxy\}.$$

(iii) Positivity: for every $x \in E$, $x \,\Box\, x \in L(E)$ is a positive operator in the sense that $i(x \,\Box\, x) \in aut(D)$ and the spectrum of $x \,\Box\, x$ is non-negative;

(iv) The C^**-condition:* For every $x \in E$

$$\| x \square x \| = \| x \|^2.$$

A Banach space E together with a map $[\,]: E \times E \to L(E)$ satisfying (i) - (iv) is called a JB^*-*triple*. Kaup proved also that *if E is a JB^*-triple then D is symmetric,* i.e. $D_s = D$ and $E_s = E$. Kaup's work [K3] goes even further, he shows that *every bounded symmetric domain in a complex Banach space E is biholomorphically equivalent to a bounded, symmetric domain which is convex and circular about the origin,* i.e. the open unit ball of an equivalent norm on E (This extends earlier results of Vigue [V1], who did not prove convexity). This establishes the equivalence of the categories of JB^*-triple systems and bounded symmetric domains with base point.

JB^*-triples generalize both C^*-algebras, JB^*-algebras (which are the complexifications of Jordan Banach algebras) and Hilbert spaces. In the case of C^*-algebra the triple product is defined by

$$\{xyz\} = (xy^*z + zy^*x)/2.$$

In the case of JB^*-algebra with Jordan product $x \circ y$, the triple product is

$$\{xyz\} = x \circ (y^* \circ z) + z \circ (y^* \circ x) - (x \circ z) \circ y^*.$$

In the case of Hilbert space with inner product (x, y) the triple product is given by

$$\{xyz\} = ((x,y)z + (z,y)x)/2.$$

Let us mention some important example of JB^*-subtriples of $L(H)$, the bounded operators on a Hilbert space H: the symmetric operators ($A^T = A$), the anti symmetric operators ($A^T = -A$). Here A^T denotes the transpose of A with respect to some orthonormal basis. Another subtriple, the spin factor, is very important in quantum mechanics but its precise description is a bit more involved.

The action of G on D in case E is a JB^*-triple is by *generalized Mobius transformations*. In case E is a C^*-algebra with a unit 1 these Mobious transformations were studied by Potapov, see [IS, chapter 8]. For every $a, z \in D$ define

$$M_a(z) = (1-aa^*)^{-\frac{1}{2}}(a-z)(1-a^*z)^{-1}(1-a^*a)^{\frac{1}{2}}.$$

One shows that $M_a \in G$ and that M_a is the symmetry which interchanges 0 and a (hence M_a fixes the midpoint along the geodesic from 0 to a). In the commutative case, i.e. $E = C(K)$ with K a compact Hausdorff space, M_a takes the familiar simple form

$$M_a(z) = (a-z)/(1-\bar{a}z); \qquad a, z \in D.$$

In case E is a Hilbert space the Mobius transformations are given in terms of the orthogonal projection P_a on Ca:

$$M_a(z) = (a - P_a z + (1 - \|a\|^2)^{1/2}(I - P_a)z)/(1 - (z,a)).$$

A recent result of Y. Friedman and B. Russo [FR3], generalizing the Gelfand- Naimark theorem for C^*-algebras, shows that the above example are not too general: Every JB^*-triple is isometrically and triple- isomorphic to a subtriple of

$$(\sum_\alpha \oplus L(H_\alpha))_\infty \oplus_\infty C(K, C^6)$$

where $\{H_\alpha\}$ are Hilbert spaces, and $C(K, C^6)$ is the space of all continuous functions on a compact Hausdorff space K with values in C^6, the exceptional JB^*- triple of all 3×3 Hermitian matrices over the Cayley numbers.

Unlike C^*-algebras, the category of JB^*-triple systems is closed under the action of contractive projections. This fundamental result is due to Y. Friedman and B. Russo [FR!], [FR2], to W. Kaup [K4] and to L. L. Stacho [S2] independently. It extends the earlier works [AF] and [ES].

5.4 Theorem: *Let E be a JB^*-triple and let $P : E \to E$ be a contractive projection. Then $P(E)$ is also a JB^*-triple when endowed with the triple product*

$$[xyz] = P(\{xyz\}); \qquad x, y, z \in P(E).$$

We present Kaup's proof.

Proof: Since $E_s = E$, Corollary 5.3 gives

$$P(E) = P(E_s) \subseteq (P(E))_s \subseteq P(E).$$

Hence $P(E) = (P(E))_s$. By Kaup's theorem mentioned above $P(E)$ is a JB^*-triple with triple product $[xyz]$. Let $a \in P(E)$ and consider

$$h_a(z) = a - \{zaz\}, \qquad z \in D$$

sp $h_a \frac{\partial}{\partial z}$ is the unique element of $aut(D)$ satisfying $h_a(0) = a$ and $h'_a(0) = 0$. Similarly, if

$$k_a(z) = a - [zaz]; \qquad z \in P(E) \cap D$$

$k_a \frac{\partial}{\partial z}$ is the unique element of $aut(P(E) \cap D)$ for which $k_a(0) = a$ and $k'_a(0) = 0$. Let

$$\tilde{h}_a = Ph_{a_{|P(E) \cap D}}.$$

Then by Theorem 5.2, $\tilde{h}_a \frac{\partial}{\partial z} \in aut(P(E) \cap D)$ and clearly $\tilde{h}_a(0) = Pa = a$ and $\tilde{h}'_a(0) = 0$. It follows that $\tilde{h}_a = k_a$. Thus

$$P\{zaz\} = [zaz]; \qquad a, z \in P(E) \cap D.$$

Polarizing, we get

$$P\{xaz\} = [xaz]; \qquad x, a, z \in P(E) \cap D.$$

□

Let us mention one more result concerning JB^*-triples. It is obvious that *the ℓ_∞-sum of a family of JB^*-triple is also a JB^*-triple,* where the triple product is defined coordinatewise. (Equivalently, the unit ball is a product of symmetric balls, hence symmetric). It follows that *the ultra product of a family of JB^*-triples is naturally a JB^*-triple.* Next, it is well-known that for every Banach space E the second dual E^{**} is naturally isometric to a 1-complemented subspace of an ultra power of E. Combining these facts with Theorem 5.4, S. Dineen proved

5.5 Theorem [D2]: *The second dual E^{**} of a JB^*-triple E is in a natural way a JB^*-triple, and E is a subtriple of E^{**} under the canonical embedding.*

Modifying the construction of Dineen, T.barton and R. M. Timoney [BT] were able to improve Theorem 5.5 by showing that *the triple product on E^{**} must be separately w^*-continuous.* Dineen [D3] generalized [D2] by showing that if $\{E_j\}$ are Banach spaces with open unit balls $\{D_j\}$, and if $h_j \frac{\partial}{\partial z} \in aut(D_j)$ are uniformly bounded then $(h_j \frac{\partial}{\partial z})$ give rise to a complete holomorphic vector field on the unit ball of the ultra product $\prod_j E_j/U$. From this he concluded Theorem 5.5 and also that *every biholomorphic automorphism of the unit ball of a complex Banach space E can be extended to a biholomorphic automorphism of the unit ball of E^{**}.*

RIENHARDT DOMAINS AND SPACES WITH 1-UNCONDITIONAL BASIS

Let E be a complex Banach space of sequences $z = (z(n))$ so that the *standard unit vectors* $\{e_k\}_{k=1}^{\infty}$ (defined via $e_k(n) = \delta_{n,k}$) form a *normalized, 1-unconditional basis for E. See [LT1], [LT2] for these and* related notations. A domain U in E is called a *Reinhardt domain* (with respect to the $\{e_k\}_{k=1}^{\infty}$) if it contains the origin and if

$$(z(n)) \in U \quad \text{if and only if} \quad (e^{i\theta_n} z(n)) \in U$$

for all choices of $\theta_n \in R, n = 1, 2, \cdots$ with $\theta_n = 0$ for all but finitely many indices n. The Reinhardt domain U is *normalized* if $e_k \in \partial U$ and $\lambda e_k \in U$ imply $|\lambda| < 1$. Clearly every Reinhardt domain is linearly equivalent to a normalized one. Finally, a convex, normalized Reinhardt domain in E is simply the open unit ball of E in an equivalent norm in which $\{e_k\}_{k=1}^{\infty}$ is still a 1-unconditional basis.

The study of Reinhardt domains depends on the following two-dimensional result of P. Thullen. Let us define for $r > 0$

$$U(r) = \{(z,w) \in C^2;\ |z|^2 + |w|^{1/r} < 1\}$$

and let

$$U(0) = \{(z,w) \in C^2;\ |z| < 1, |w| < 1\}.$$

5.6 Theorem [T]. *Let U be a bounded, normalized Reinhardt domain in C^2.*

(i) Either $Aut(U)$ consists of linear maps, or $U = U(r)$ for a unique $r \in [0,\infty)$, up to a permutation of the coordinates;

(ii) $U(0)$ and $U(\frac{1}{2})$ are symmetric, being the open unit balls of the 2-dimensional ℓ_∞ and ℓ_2 spaces respectively;

(iii) For $0 < r, r \neq \frac{1}{2}$, $Aut(U(r))$ consists of all maps

$$\Phi(z,w) = (\varphi(z), \lambda(\varphi'(z))^r w)$$

with $\varphi \in Aut(\Delta)$ and $|\lambda| = 1$. In particular the symmetric part of $U(r)$ is

$$U(r)_s = Aut(U(r))(0) = \Delta \times \{0\}.$$

This classical theorem shows in particular that the Riemann mapping theorem is not valid in dimension n for $n > 1$. For instance

5.7 Corollary: *The bidisc $U(0) = \Delta \times \Delta$ and the unit ball $U(\frac{1}{2})$ in C^2 are not biholomorphically equivalent.*

5.8 Corollary: *for $p \neq 2,\infty$ all biholomorphic automorphisms of*

$$\{(z,w) \in C^2;\ |z|^p + |w|^p < 1\}$$

are linear. In particular this domain is not biholomorphically equivalent to $U(0)$ or $U(\frac{1}{2})$.

Theorem 5.6 was extended by T. Sunada [SU] to C^n and then by L. L. Stacho [S2], J. P. Vigue [V5] and T. Barton [B1], [B2] to the infinite dimensional setting. See also [BDT] for this and for the study of Reinhardt decompositions of Banach spaces. To describe these results we adopt the notation

$$E_A = \overline{span}\ \{e_k\}_{k \in A},\ U_A = U \cap E_A$$

for every $A \subseteq N$.

5.9 Theorem [BDT]: *Let U be a bounded normalized Reinhardt domain in E.*

(i) There exists a subset I of the positive integer N with complement $J = N \setminus J$ and a partition $\mathcal{P}$ of I so that $E_s = E_I$ and so that the sequence $\{e_i\}_{i \in I}$ is isometrically equivalent to the unit vector basis of $(\sum_{p \in \mathcal{P}} \oplus \ell_2(p))_{c_0}$;

(ii) $(Aut(U))(0) = U_I$;

(iii) There exist non-negative numbers $\{r_{p,j} \, ; p \in \mathcal{P}, j \in J\}$ satisfying

$$\sup_{j \in J} \sum_{p \in \mathcal{P}} r_{p,j} < \infty \tag{*}$$

so that $(z(n)) \in U$ if and only if

(a) $(z_p)_{p \in \mathcal{P}} \in c_0(\mathcal{P})$ and $\max_{p \in \mathcal{P}} z_p < 1$, where $z_p = (\sum_{i \in \mathcal{P}} |z(i)|^2)^{1/2}, p \in \mathcal{P}$

(b) $\sum_{j \in J} (\prod_{p \in \mathcal{P}} (1 - z_p^2)^{-r_{p,j}}) z(j) e_j \in U_J$.

Moreover, if U is convex then U_J is convex and

$$\sup_{j \in J} \sum_{p \in \mathcal{P}} r_{p,j} \leq 1. \tag{**}$$

5.10 Remarks: (i) The first part of Theorem 5.9 concerning general Reinhardt domains admits a converse. Given $I, J, \mathcal{P}$ and $\{r_{p,j}\}$ satisfying (*) there exists a Reinhardt domain U consisting of all $(z(n))$ so that the conditions (a) and (b) are satisfied. For the convexity part of Theorem 5.9 - it is not true that in general the convexity of U_J and (**) imply the convexity of U. See [B1] and [BDT] for these matters.

(ii) If $0 < r < \infty, r \neq 1/2$, then $(z, w) \in U(r)$ if and only if

$$|z| < 1 \quad \text{and} \quad \frac{|w|}{(1 - |z|^2)^r} < 1.$$

So, in the notation of Theorem 5.9, $I = \{1\}$, $\mathcal{P} = \{I\}$, $J = \{2\}$, $r_{I,2} = r$ and $U_J = \{0\} \times \Delta$.

(iii) The subspaces $E_p, p \in \mathcal{P}$ of E are known as *Hilbert components* of E. See [KW].

5.11 Corollary: *E is symmetric (in the sense that $E = E_s$) if and only if E is the c_0-sum of a sequence of Hilberts spaces: In particular, if E is a symmetric sequence space (i.e. the unit vector basis $\{e_k\}_{k=1}^{\infty}$ form a 1-symmetric basis of E) then either $E_s = \{0\}$ or $E_s = E$. In the last case, either $E = \ell_2$ or $E = c_0$.*

SPACES WITH THE LINEAR BIHOLOMORPHIC PROPERTY

A complex Banach space E with open unit disc D is said to have the *linear biholomorphic property* (*LBP*, for short) if $G = Aut(D)$ consists of linear operators. Equivalently, if $\underline{g} = aut(D)$ consists of linear operators. *LBP* is the extreme opposite to symmetricity since E has *LBP* if and only if its symmetric part is trivial, namely $E_s = \{0\}$.

The main tool in studying *LBP* in Banach spaces is the following.

5.12 Proposition [S2]: *Suppose $\mathcal{P}$ is a family of contractive projections in E, satisfying*

(i) $P(E)$ has LBP for every $P \in \mathcal{P}$

(ii) $\bigcap_{P \in \mathcal{P}} \ker P = \{0\}$.

Then E has LBP.

Proof: For every $P \in \mathcal{P}$ we have by (i) and Corollary 5.3

$$P(E_s) \subseteq (P(E))_s = \{0\}.$$

Thus $E_s \subseteq \ker P$ for every $P \in \mathcal{P}$ Therefore $E_s = \{0\}$ by (ii).

□

5.13 Corollary ([S1],[S2],[BKU]): *Let E be a rearrangement-invariant Banach function space on either N,[0,1] or $[0,\infty)$. If E is not a Hilbert space or a C^*- algebra then E has LBP.*

For the study of rearrangement-invariant spaces see [LT2]. If the measure space is N then E is a symmetric sequence space and the assertion follows from Corollary 5.11. For the other two cases the assertion follows from Proposition 5.1 by taking $\mathcal{P}$ to be the family of all conditional expectations with respect to finite σ-algebras, generated by finite families of disjoint measurable sets with the same measure.

5.14 Corollary: *[A] Let E be a symmetric sequence space different from ℓ_2 and c_0 and let S_E be the associated unitary ideal of operators on ℓ_2. Then S_E has LBP.*

S_E is defined as the Banach space of all operators T on ℓ_2 whose sequence of singular numbers $(s_n(T))$ belongs to E, normed by

$$\|T\|_{S_E} = \|(s_n(T))\|_E .$$

See [GK]. Corollary 5.14 follows from Proposition 5.12 and Corollary 5.13 by taking $\mathcal{P}$ to be the family of all generalized diagonal projections P of the form

$$P(T) = \sum_{n \in I} (Th_n, f_n)(\cdot, h_n) f_n$$

where $\{h_n\}_{n \in I}$, $\{f_n\}_{n \in I}$ are (finite or infinite) orthonormal sequences. The point is that the sequence, $\{(\cdot, h_n) f_n\}_{n \in I}$ is isometrically equivalent to the sequence $\{e_n\}_{n \in I}$ in E, and so $P(S_E)$ has *LBP* if $|I| \geq 2$.

Let $H_1, \ldots, H_n$ be Hilbert spaces with $\dim(H_j) \geq 2$. Let $H_1 \hat{\otimes} H_2 \hat{\otimes} \cdots \hat{\otimes} H_n$ be the injective tensor product, i.e. the space of all bounded n-linear maps $f : H_1 \times H_2 \times \cdots \times H_n \to C$, normed by $\|f\| = \sup\{|f(z_1, \ldots, z_n)|;\ z_j \in H_j,\ \|z_j\| \leq 1\}$. If $n = 2$, $H_1 \hat{\otimes} H_2$ is identified with the space of all compact operators from H_1 into H_2 which is a JB^*-subtriple of $L(H_1 \oplus H_2)$. For $n > 2$ the situation changes drastically

5.15 Proposition [S2]: *For $n > 2$ the space $H_1 \hat{\otimes} H_2 \hat{\otimes} \cdots \hat{\otimes} H_n$ has LBP.*

In an another direction we have

5.16 Proposition [BKU]: *for $1 \leq p < \infty, p \neq 2$, the Hardy spaces H^p have LBP.*

UNIFORM ALGEBRAS

A *uniform algebra* A is a closed subalgebra of $C(K)$, K a compact Hausdroff space, which contains the unit 1 and separates the points of K. A is endowed with the supremum norm.

5.17 Proposition [BKU]: *Let A be a uniform algebra. Then the symmetric part of A is its maximal C^*-subalgebra, i.e.*

$$A_s = \{ f \in A ; \overline{f} \in A \}.$$

From this it is elementary to see

5.18 Corollary [BKU]: *Let A be either the disc algebra or H^∞ over the unit disc Δ. Then $A_s = C1$.*

E_s NEED NOT BE COMPLEMENTED IN E

A natural question concerning the symmetric part E_s of a Banach space E is whether E_s must be complemented in E. The following unpublished example of J. P. Vigue answers this negatively.

5.19 Proposition: *There exists an equivalent* 1-*symmetric norm* $|||\cdot|||$ *on* ℓ_∞ *so that if* $E = (\ell_\infty, |||\cdot|||)$ *then* $E_s = c_0$. *In particular* E_s *is not isomorphic to a complemented subspace of* E.

Indeed, fix $0 < \varepsilon < 1$ and let D_0 and D_∞ be the open unit balls of c_0 respectively ℓ_∞. Define

$$D = (1-\varepsilon)D_0 + \varepsilon D_\infty.$$

Then D is an open, convex subset of D_∞, $\varepsilon D_\infty \subset D$ and D is invariant under permutations and changes of complex signs of the coordinates. Thus D is the open unit ball of ℓ_∞ in an equivalent, 1-symmetric norm $|||\cdot|||$. Set $E = (\ell_\infty, |||\cdot|||)$. Let $G = Aut(D)$, and let G^0 be the connected component of the identity in G. We claim that

$$G^0(0) = D_0.$$

Clearly, having this the proposition follows from Lemma 4.2 and the fact that $D_0 = D \cap D_0$. To prove the claim, observe first that

$$D = \{z = (z_n) \in D_\infty ; \limsup_{n\to\infty} |z_n| < \varepsilon\}.$$

Next, consider the maps of the form

$$\varphi(z_1, z_2, \ldots, z_n, \cdots) = (\varphi_1(z_1), \varphi_2(z_2), \ldots, \varphi_n(z_n), \ldots) \qquad (*)$$

where $z_j \in \Delta$ and $\varphi_j \in Aut(\Delta)$. It is elementary to verify that such φ is a member of $Aut(D)$ if and only if $\varphi(0) = (\varphi_j(0)) \in D_0$. Let $\tilde{G}$ denotes the set of these maps (*) with $\varphi(0) \in D_0$. Then $\tilde{G}$ is a connected subgroup of $Aut(D)$. The main point is that $\tilde{G}$ is open in $Aut(D)$, see [V3, Th. 1.8]. But an open subgroup of a topological group is always closed. Thus $\tilde{G}$ is open, closed and connected. Hence $\tilde{G} = G^0$. Thus the symmetric part of D is

$$G^0(0) = \tilde{G}(0) = D_0.$$

□

REFERENCES

[A] Arazy, J., On biholomorphic automorphisms of the unit ball of unitary matrix spaces, Linear Algebra and its Applications, 80(1986), 180-182.

[AF] Arazy, J.and Friedman,Y., Contractive projections in C_1 and C_∞, Memoirs of Amer. Math. Soc., 13 (1978), No. 200.

[B1] Barton, T., Bounded Reinhardt domains in complex Banach spaces, Ph.D. Thesis, Kent State University (1984).

[B2] Barton, T. Biholmorphic equivalence of bounded Reinhardt domains, An. Sci. Norm. Sup, Pisa (to appear).

[BDT]
Barton, T., Dineen, S. and Timoney, R., Bounded Reinhardt domains in Banach spaces, Compositio Mathematica (to appear).

[BT] Barton, T., and Timoney, R., Weak * continuity of Jordan triple products and applications, Math. Scand. (to appear).

[BKU]
Braun, R., Kaup, W. and Upmeier, H., On the automorphisms of circular and Reinhardt domains in complex Banach spaces, Manuscripta Math. 25 (1978), 97-133.

[CE] Cartan, E., Sur les domaines bornés homogènes de l'espace de n variables complexes, Abh. Math. Sem. Univ. Hamburg 11(1935), 116-162.

[CH1]
Cartan, H., Les fonctions delta de deux variables complexes et le problème de la représentation analytique, J. Math. Pures. Appl. 9e série, 10 (1931), 1-144.

[Ch2]
Cartan, H., Sur les groupes de transformations analytiques, Act. Sci. Indust., Exposés Math., 9, Hermann, Paris, 1935.

[D1] Dineen, S., Complex analysis in locally convex spaces, North Holland Math. Studies, 57, 1981.

[D2] Dineen, S., The second dual of a JB^*-triple system. Complex analysis, functional analysis and approximation theory, J. Mujica (editor), North. Holland, Ud. 125, 1986.

[D3] Dineen, S., Complete holomorphic vector fields on the second dual of a Banach space, Math. Scand. (to appear).

[ES] Effros, E. and Størmer, E., Positive projections and Jordan structure in operator algebra, Math. Scand. 45 (1979), 127-138.

[FV] Franzoni, T. and Vesentini, E., Holomorphic maps and invariant distances, North Holland Math. Studies, 40, 1980.

[FR1]
Friedman, Y. and Russo, B., Contractive projections on operator triple systems, Math. Scand. 52 (1983), 279-311.

[FR2]
Friedman, Y. and Russo, B., Solution of the contractive projection problem, J. Funct. Anal. 60 (1985), 56-79.

[FR3]
Friedman, Y. and Russo, B., The Gelfand Naimark theorem for JB^*-triples, Duke Math. J.

(to appear).

[GK] Gohberg, I.C. and Krein, M.G., Introduction to the theory of linear non- self adjoint operators, Amer. Math. Soc. Translations, Vol. 18.

[H1] Harris, L., Bounded symmetric homogeneous domains in infinite dimensional spaces, Lecture notes in Math. 364, Springer-Verlag, New York-Heidelberg- Berlin, 1974, 13-40.

[H2] Harris, L., Schwarz Pick systems of pseudometries for domains in normed linear spaces, Advances in Holomorphy, J. A. Barroso (ed.), North Holland, 1979, 345-406.

[H3] Harris, L., A generalization of C^*-algebras, Proc. London. Math. Soc. (3) 42 (1981), 331-361.

[IS] Isidro, J. M. and Stacho, L. L., Holomorphic automorphism groups in Banach spaces: an elementary introduction, North Holland Math. Studies, 105 (1985).

[IV] Isidro, J. M. and Vigué, J.P., Sur la topologie du groupes des automorphismes analytiques d'un domaine cercle borné, B.S.M.F. 106 (1982), 417-426.

[KW] Kalton, N. J. and Wood, G.V., Orthonormal systems in Banach spaces and their applications, Math. Proc. Camb. Phil. Soc. 79 (1976), 493-510.

[K1] Kaup, W., Algebraic characterization of symmetric complex Banach manifolds, Math. Ann. 228 (1977), 39-64.

[K2] Kaup, W., Jordan algebras and holomorphy, Functional analysis, holomorphy and approximation theory, Lecture Notes in Mathematics 843, Springer-Verlag, Berlin-Heidelberg-New York, 1981, 341-365.

[K3] Kaup, W., A Riemann mapping theorem for bounded symmetric domains in complex Banach spaces, Math. Z. 183 (1983) 503-529.

[K4] Kaup, W., Contractive projections on Jordan C^*-algebras and generalizations, Math. Scand. 54 (1984), 95-100.

[KU] Kaup, W. and Upmeier, H., Banach spaces with biholomorphically equivalent unit balls are isomorphic, Proc. Amer. Math. Soc. 58 (1978), 129-133.

[LT1] Lindenstrauss, J. and Tzafriri, L., Classical Banach spaces I. Ergebnisse 92, Springer-Verlag, New York-Heidelberg-Berlin, 1977.

[LT2] Lindenstrauss, J. and Tzafriri, L., Classical Banach spaces II. Ergebnisse 97, Springer-Verlag, New York-Heidelberg-Berlin, 1979.

[L] Loos, O., Bounded symmetric domains and Jordan pairs, Lecture Notes, University of California at Irvine, 1977.

[R1] Rosenthal, H., The Lie algebra of a Banach space, preprint.

[R2] Rosenthal, H., Functional Hilbertian sums, preprint.

[S1] Stacho, L. L., A short proof that the biholomorphic automorphisms of the unit ball of L^p spaces are linear, Acta. Sci. Math. 41 (1979), 381-383.

[S2] Stacho, L. L., A projection principle concerning biholomorphic automorphisms, Acta. Sci. Math. 44 (1982), 99-124.

[SU] Sunada, T., Holomorphic equivalence problem for bounded Reinhardt domains, Math. Ann. 235 (1978), 111-128.

[T] Thullen, P.,Die Invarianz des Mittelpunktes von Kreiskörpern, Math. Ann. 104 (1931) 244-259.

[U1] Upmeier, H., Über die Automorphismengruppen beschränkter Gebiete in Banachräumen. Dissertation, Tübingen, 1975.

[U2] Upmeier, H., Über die Automorphismengruppen von Banach-Mannigfaltigkeiten mit invarianter Metrik, Math. Ann. 223 (1976), 279-288.

[U3] Upmeier,H., Symmetric Banach Manifolds and Jordan C^*-algebras, North Holland Math. Studies, 104 (1985).

[U4] Upmeier, H., Jordan algebras in analysis, operator theory and quantum mechanics, CBMS-NSF Regional Conference, Irvine, 1985.

[V1] Vigué, J. P., Le groupe des automorphismes analytiques d'un domaine borné d'un espace de Banach complexe. Applications aux domaines bornés symétriques, Ann. Sci. Ec. Norn. Sup. 4e série 9 (1976), 203-282.

[V2] Vigué, J. P., Les domaines bornés symétriques d'un espace de Banach complexe et les systèmes triples de Jordan, Math. Ann. 229 (1977), 223-231.

[V3] Vigué, J. P., Automorphismes analytiques des produits de domaines bornés, Ann. Sci. Ec.Norm. Sup. 4e série 11 (1978), 229-246.

[V4] Vigué, J. P., Les automorphismes analytiques isométriques d'une variété complexe normée, Bull.Soc. Math. France 110 (1982), 49-73.

[V5] Vigué, J. P., Automorphismes analytiques d'un domaine de Reinhardt borné d' un espace de Banach à base, Ann. de l'Inst. Fourier (Grenoble), 34,2 (1984), 67-87.

A DENSITY CONDITION FOR ANALYTICITY OF THE RESTRICTION ALGEBRA

J. BOURGAIN
IHES, France and University of Illinois

1. INTRODUCTION

This paper has to be considered as an appendix to [1] (see especially remark (2) at the end). More precisely, we will use the method developed in [1] to obtain a significant improvement of the density criterion for analyticity of the restriction algebra proved by Katznelson and Malliavin in [4]. A rather detailed discussion of their results can also be found in the book of Graham and McGehee [3] (see Open Problem section). [3] is also one standard reference work for background material. Next I recall some definitions and facts. Let G be a compact Abelian group, $\Gamma = \hat{G}$ the dual group of G. For $\Lambda \subset \Gamma$, let

$$A(\Lambda) = \{\hat{f}\mid_{\Lambda}; f \in L^1(G)\}$$

be the restriction algebra (of Fourier transforms of L^1 - functions on G).

The set $\Lambda \subset \Gamma$ is dissociated if any ($\pm 1,0$) relation on the characters of Λ is trivial, thus

$$\Sigma_{\Lambda}\varepsilon_{\gamma}\gamma = 0,\ \varepsilon_{\gamma} = 1,-1,0 \Rightarrow \varepsilon_{\gamma} = 0 \quad \text{for} \quad \gamma \neq 0.$$

The set $\Lambda \subset \Gamma$ is called a Sidon set provided there is a constant $C > 0$ such that

$$C \,\|\Sigma_{\Lambda} a_{\gamma}\gamma\|_{C(G)} \geq \Sigma |a_{\gamma}| \tag{1}$$

holds for all scalar sequences $(a_{\gamma})_{\gamma\in\Lambda}$. The smallest constant C satisfying inequality (1) is called the Sidon constant $S(\Lambda)$ of Λ. Here, $\mathcal{C}(G)$ refers to the space of continuous functions on G endowed with the uniform norm and characters γ are considered as functions on the group.

Dissociated sets are special cases of Sidon sets. The interpolating measures are then given by the standard Riesz products.

The algebra $A(\Lambda)$ is analytic provided only analytic functions operate on $A(\Lambda)$. Recall that $F:[-1,1]\to C$ operates on $A(\Lambda)$ provided

$$\varphi \in A(\Lambda), \varphi(\gamma) \in]-1,1[\Rightarrow F \circ \varphi \in A(\Lambda)$$

It is known that for $A(\Lambda)$ to be analytic, it is necessary and sufficient that for some $c > 0$

$$N(\Lambda,t) \equiv \sup_{\varphi \text{ real}, \|\varphi\|_A \leq 1} \|e^{it\varphi}\|_A \geq e^{ct}. \tag{2}$$

for all $t > 0$ large enough. Here of course $\|\ \|_A$ refers to the $A(\Lambda)$ - norm (= quotient norm) , i.e.

$$\|\varphi\|_A = \inf\{\|f\|_{L^1(G)}; \hat{f}(\gamma) = \varphi(\gamma) \quad \text{for} \quad \gamma \in \Lambda\}.$$

We consider here the following version of the dichotomy problem .

Conjecture (*) : Either Λ is a Sidon set or any function operating on $A(\Lambda)$ is analytic.

Graham pointed out that the next fact [4] is a consequence of (*).

THEOREM 1 : If $B(\Lambda) = B_d(\Lambda)$, i.e. the restriction algebra's of measures and discrete measures coincide, then Λ is a Sidon set and hence an I -set

Another support for (*) is its affirmative solution in the tensor - algebra case [1].

THEOREM 2 : Let I, J be discrete spaces and $E \subset I \times J$. Then either E is a V-Sidon set or the restriction algebra $A(E) = \ell^\infty(I)\hat{\otimes}\ell^\infty(J)/E^\perp$ is analytic.

Conjecture (*) has not been settled for any infinite group G. In [2] a density condition in Λ is given which asserts the analyticity of $A(\Lambda)$. Let us make following definition

$$\rho_\Lambda(k) = \inf_{\substack{E \subset \Lambda \\ |E| = k}} \sup\{ |F|; F \subset E \text{ and } F \text{ dissociated "}\}".$$

For Λ to be a Sidon set, it is necessary and sufficient that

$$\inf_k \frac{S_\Lambda(k)}{k} > 0 \tag{3}$$

Let G be a bounded group, i.e. which elements are of bounded order. For example $G = \{1,-1\}^N$, the Cantor group. Then [2] assests analyticity of $A(\Lambda)$ provided

$$\underline{\lim} \frac{\rho_\Lambda(k)}{\log k} < \infty \tag{4}$$

Clearly there is a large gap between (3) and (4). I will prove here the following dichotomy.

PROPOSITION 3 : Let G be a bounded compact Abelian group and $\Lambda \subset \Gamma$. Then either $A(\Lambda)$ is analytic or for some $\delta > 0$

$$\rho_\Lambda(k) \geq k^\delta, \forall k \tag{5}$$

The proof is related to [1].

2. Proof of Proposition 3 Recall the $\| e^{it\phi} \|_{A(E)}$ minoration proved in [1] (section 2) .

LEMMA 4 : Let G be compact Abelian and $\Lambda \subset \Gamma$. Suppose there exists f with supp $\hat{f}$ finite and contained in Λ, as well as points $x_1, \ldots, x_\ell$ in G such that

$$f(0) = \| f \|_\infty = 1 \tag{6}$$

$$\sum_{S\subset\{1,\ldots,\ell\}} |f(y-\sum_{k\in S} x_k)| \le B \quad \text{for} \quad y\in G \tag{7}$$

Then

$$N(\Lambda,t) \ge e^{ct} \quad \text{for} \quad B \le t \le \ell$$

where $c > 0$ is numerical.

This lemma is then applied by taking a finite $E \subset \Lambda$ and defining

$$f = \frac{1}{|E|}\sum_{\gamma\in E}\gamma$$

Condition (7) is verified in the mean. Thus we evaluate

$$\int_{G\times\cdots\times G} \| \sum_{S\subset\{1,\ldots,\ell\}} |f_{\sum_{k\in S} x_k}| \|_\infty dx_1 \ldots dx_\ell \tag{8}$$

where for $z \in G$, f_z stands for the translate of f by z. The dychotomy expressed in Proposition 3 amounts to proving that either $A(\Lambda)$ is analytic or there is a constant M satisfying the condition

$$|gr(I) \cap \Lambda| \le |I|^M \tag{9}$$

wherever I is a finite subset of Λ. Here $gr(I)$ is the group generated by I. By hypothesis, $|gr(I)| \le C(G)^{|I|}$. Assume (9) does not hold for given M (fixed). Thus there is $I \subset \Lambda, |I| = N$ and $|gr(I) \cap \Lambda| > N^M$ (Here N can be taken arbitrarily large).

Define for integers $k > 0$.

$$\phi(k) = \sup\{ |gr(E) \cap \Lambda| ; E = \Lambda \text{ and } |E| \subset k\} \tag{10}$$

Thus from the preceding

$$\phi(N) > N^M \tag{11}$$

Therefore, there exists an integer $1 < n < \frac{N}{4}$ for which

$$\phi(4n) > 2^M \phi(2n) \tag{12}$$

Let $E \subset \Lambda, |E| = 4n$ for which $F = \Lambda \cap gr(E)$ satisfies

$$|F| = \phi(4n) \tag{13}$$

Denote $g = \sum_{\gamma\in F} \gamma$ and evaluate for fixed ℓ (to be specified later)

$$\int_{G^\ell} \| \sum_{S\subset\{1,\dots,\ell\}} | g_{\sum_{k\in S} x_k} | \|_\infty dx_1 \cdots dx_\ell \tag{14}$$

If $2^\ell < n$, then for all $x_1,\dots,x_\ell \in G$

$$\| \Sigma_S | g_{\sum_{k\in S} x_k} | \|_\infty \le$$

$$2 \sup_{a_S = \pm 1} \| \Sigma_s a_s g_{\sum_{k\in S} x_k} \|_\infty \le$$

$$2C(G) \sup_{a_S = \pm 1} \| \Sigma_S a_S g_{\sum_{k\in S} x_k} \|_{2n}$$

$$\le 2\sqrt{2} C(G) \sup_{a_S = \pm 1} \| \Sigma_S a_S g(x + \sum_{k\in S} x_k) \|_{L^{2n}(x, x_1, \dots, x_\ell)} \tag{15}$$

Here $C(G)$ is a constant depending on the (bounded) group G.

Rewriting

$$\Sigma_S a_S g(x + \sum_{k\in S} x_k) = \Sigma_S \sum_{\gamma\in F} a_S \gamma(x) \prod_{k\in S} \gamma(x_k)$$

it is clear that

$$\| \Sigma_S a_S g(x + \sum_{k\in S} x_k) \|_{L^{2n}(x, x_1, \dots, x_\ell)} \le \| \Sigma_S \sum_{\gamma\in F} \prod_{k\in S} \gamma(x_k) \|_{L^{2n}(x_1, \dots, x_\ell)} \tag{16}$$

We have

$$\| \Sigma_S \sum_{\gamma\in F} \prod_{k\in S} \gamma(x_k) \|_{2n}^{2n} = \Sigma_{F^{2n}} c(\gamma_1, \dots, \gamma_{2n})^\ell \tag{17}$$

where

$$c(\gamma_1, \dots, \gamma_{2n}) = \int_G [1 + \gamma_1(x)] \cdots [1 + \gamma_n(x)][1 + \gamma_{n+1}(-x)] \cdots [1 + \gamma_{2n}(-x)] dx$$

Define

$$D(\gamma_1, \dots, \gamma_{2n}) = \{1 < j \le 2n \mid \gamma_j \in gr(\gamma_1, \dots, \gamma_{j-1})\}$$

$$d(\gamma_1, \dots, \gamma_{2n}) = | D(\gamma_1, \dots, \gamma_{2n}) |$$

Then

$$c(\gamma_1,...,\gamma_{2n}) = \int_G \prod_{\substack{1\le j\le n \\ j\notin D}} [1+\gamma_j(x)] \prod_{\substack{n<j\le 2n \\ j\notin D}} [1+\gamma_j(-x)] \prod_{j\in D} [1+\gamma_j(\pm x)]dx$$

can be estimated by

$$2^{d(\gamma_1,...,\gamma_{2n})} \| \prod_{\substack{1\le j\le 2n \\ j\notin D}} [1+\gamma_j(\pm x)] \|_{PM(\Gamma)} = 2^{d(\gamma_1,...,\gamma_{2n})}$$

Consequently, by (15), (16), (17)

$$\le 2\sqrt{2}\, C(G)[\Sigma_{F^{2n}} 2^{\ell\cdot d(\gamma_1,\cdots\gamma_{2n})}]^{1/2n} \tag{18}$$

By definition of $D(\gamma_1,...,\gamma_{2n})$ and (10), (12) , (13)

$$\#\{(\gamma_1,...,\gamma_{2n}) \in F^{2n}) \mid d(\gamma_1,....,\gamma_{2n} = k\} \le$$

$$\binom{2n}{k} |F|^{2n-k} \Big(\sup_{\substack{|E|\le 2n-1 \\ E\subset F}} |F \cap gr(E)| \Big)^k \le$$

$$\binom{2n}{k} \Big(\frac{\phi(2n)}{|F|}\Big)^k |F|^{2n} =$$

$$\binom{2n}{k} \Big(\frac{\phi(2n)}{\phi(4n)}\Big)^k |F|^{2n} \le \binom{2n}{k} 2^{-Mk} |F|^{2n}. \tag{19}$$

Substitution of (19) in (18) yields

$$\Sigma_{F^{2n}} 2^{\ell d(\gamma_1\cdots\gamma_{2n})} \le \sum_{k=0}^{2n} 2^{\ell k} 2^{-Mk} \binom{2n}{k} |F|^{2n} = (1+2^{\ell-M})^{2n} |F|^{2n}$$

$$(14) \le 2\sqrt{2} C(G)(1+2^{(\ell-M)}) |F| \tag{20}$$

For fixed $t>0$ and c as in Lemma 4, let $\ell > \frac{t}{c}$ and take $M=\ell$

If $f = \frac{1}{|F|}\Sigma_{\gamma\in F}\gamma$, F the subset of Λ obtained above, it results from (20) that expression (8) is dominated by $8\, C(G)$.

Using Lemma 4, the minoration

$$N(\Lambda,t) \geq \frac{1}{8C(G)} C^{ct} \tag{21}$$

then follows. This inequality can thus be obtained for arbitrarily large t provided (9) is assumed to fail for M taken large enough. This ends the proof.

References

[1] J. BOURGAIN: On the dichotomy problem for tensor algebras, TAMS, Vol 293, N2, 1986, 793 - 798

[2] J. BOURGAIN: Sur les ensembles d'interpolation pour les mesures discrètes CR Acad Sci Paris, t.296, Sér I, 149-151 (1982)

[3] C. GRAHAM, O.C. McGEHEE: Essays in commutative harmonic analysis, Springer N - Y, 1982

[4] Y. KATZNELSON, P. MALLIAVIN: Un critère d'analyticité pour les algèbres de restriction, CR Acad Sci Paris, Sér A -3 261, A4964-A4967 (1965)

REMARKS ON THE EXTENSION OF LIPSCHITZ MAPS DEFINED ON DISCRETE SETS AND UNIFORM HOMEOMORPHISMS

J. BOURGAIN

IHES, France and University of Illinois

1. INTRODUCTION

Our main reference is the survey paper [Benj]. If $F : X \to Y$ is a uniform homeomorphism between the Banach spaces X, Y, then there is a constant K satisfying

$$(1) \qquad K^{-1} \le \frac{\|F(x)-F(x')\|}{\|x-x'\|} \le K \quad \text{if} \quad \|x-x'\| \ge 1.$$

Thus F becomes a bi-Lipschitz map onto its image when restricted to a discrete net in X of 1-separated (or more generally δ-separated) points. This remark relates the two concepts appearing in title.

A theorem of Ribe asserts that if X, Y are uniformly homeomorphic then they have the same local structure. In fact, if there is a Lipschitz embedding of a 1-net of points in X into Y, then X is finitely representable in Y. Let E be an n-dimensional normed space. Then there is $\delta > 0$ such that if $\mathcal{E}_\delta$ is a δ-set in the unit sphere of E and f a Lipshitz embedding of $\mathcal{E}_\delta$ into a normed space Y, then there is a subspace F of Y satisfying

$$\dim F = n \qquad d(E,F) \le 2\, dist\, f$$

where $d(\cdot,\cdot)$ is the Banach Mazur distance and $dist\, f$ is the distortion of f, i.e. $\|f\|_{lip} \|f^{-1}\|_{lip}$. In what follows, we will give a new proof of this fact which also yields an estimate on δ.The first step consists in replacing f by a Lipshitz map $\bar{f}$ defined on the whole space E. More precisely

Theorem 1: Let $S = [\|x\| = 1]$ be the unit sphere of E, $\dim E = n$ and $\mathcal{E}_\delta$ a δ-net in S. Let $f : \mathcal{E}_\delta \to Y$ have Lipschitz constant L. Let $\tau > C\delta$. Then there is a map $\bar{f} : E \to Y$ satisfying

$$\|f(x) - \bar{f}(x)\| \le \tau L \quad \text{for } x \in \mathcal{E}_\delta \tag{2}$$

$$\|\bar{f}\|_{lip} \le C(1+\delta\tau^{-1}n)L \tag{3}$$

(C is numerical).

Thus in order to bound $\|\bar{f}\|_{lip}$, one has to choose $\delta < \frac{1}{n}$. As will be clear later, this condition is reasonable. By suitable convolution (E is finite dimensional), the map $\bar{f}$ considered above can be assumed to be smooth. The next step consists in differentiating $\bar{f}$ at some point in order to get a linear into- isomorphism from E into Y. We achieved this for some $\delta = \delta(n)$, however considerably smaller than $\frac{1}{n}$ ($\log\log \frac{1}{\delta(n)} \sim n$).

Assume the map f in Theorem 1 a Lipschitz one-to-one map of $\mathcal{E}_\delta$ into Y with a bounded distorsion. If τ is taken small enough and $\delta = \frac{\tau}{n}$, the map $\bar{f}$ will have bounded Lipschitz constant and separate points in S sufficiently far apart. The existence of such a map $\bar{f}$ may already have considerable implications.

Theorem 2: Let $\mathcal{E}$ be a $\frac{1}{n}$-net in the euclidean sphere $S_{n-1} = [(\sum_{i=1}^{n} x_i^2)^{1/2} = |x| = 1]$ and $f : \mathcal{E} - Y$ a one-to-one map. There is a subspace H of Y, H 2-Hilbertian and

$$\dim H \geq cn\,(dist\ f)^{-4} \tag{4}$$

Let $\delta > 0$, $\mathcal{E}_\delta$ a δ-net in S_{n-1} and $Y = \ell_n^p$. The Mazur-map

$$f(x) = \{|x_i|^{2/p-1} x_i\}_{i=1}^n$$

is easily seen to satisfy a bound on $dist\,(f \mid \mathcal{E}_\delta)$ provided $\delta^{-|p-2|}$ is bounded. On the other hand, if $p > 2$, ℓ_n^p only contains a proportional Hilbertian subspace provided $\frac{1}{p-2} \gtrsim \log n$. This examples shows that the statement in Theorem 2 requires Lipschitz embedding of an $\mathcal{E}_\delta$-net where δ has to satisfy at least the condition

$$\log \frac{1}{\delta} > c \log n. \tag{5}$$

It follows from Ribe's theorem that the metric structure of a Banach space determines its finite dimensional structure. The converse is not true. For instance, ℓ^1 and $L^1[0,1]$ are not uniformly homeomorphic, as proved by P. Enflo.The analogue question for the spaces ℓ^p and $L^p[0,1]$, $(1 < p < \infty, p \neq 2)$ is open already for some time. We solve the problem for $1 < p < 2$ by showing the following fact.

Theorem 3: No map $F : \ell^2 \to \ell^p$ $(1 \leq p < 2)$ fulfills property (1) stated above.

The argument seems not adaptable to the case $p > 2$. Let us recall at this point that by a result of I. Aharoni every separable space can be Lipschitz embedded in the sequence space c_0.

2. PROOF OF THEOREM 1

Let E be a real n-dimensional normed space with unit ball $B = \{x \in \Re^n; \|x\| \leq 1\}$. Assume the n-dimensional volume $Vol\ B = 1$. Let $\mathcal{E}$ be a δ-net in the unit sphere $S = \partial B$. Let X be the indicator function of B and define $\chi_t(x) = t^{-n}\chi(t^{-1}x)$ $(t > 0)$. The infinite convolution

$$K = \chi_{2^{-1}} * \chi_{2^{-2}} * \chi_{2^{-3}} * \cdots$$

has the following properties

$$K \geq 0, \int_{\Re^n} K(x)dx = 1 \tag{6}$$

$$supp\ K \subset B \quad \text{(convexity)} \tag{7}$$

$$K = \chi_{1/2} * K_{1/2}. \tag{8}$$

Consider a system $(\varphi_p)_{p \in \mathcal{E}}$ of smooth [0,1]-valued functions on $\Re^n$ satisfying

$$\varphi_p(x) = 0 \ \text{ if } \ \|x-p\| > 2\delta \ (p \in \mathcal{E}) \tag{9}$$

$$\Sigma_{\mathcal{E}}\varphi_p(x) = 1 \quad \text{for } \ x \in S. \tag{10}$$

To achieve this, it suffices to take for each $p \in \mathcal{E}$ a smooth [0,1]-valued function ψ_p with $\psi_p(x) = 1$ if $\|x-p\| \leq \delta$ and $\psi_p(x) = 0$ for $\|x-p\| > 2\delta$. Since $\mathcal{E}$ is δ-dense in S, $\prod_{p \in \mathcal{E}} (1-\psi_p) = 0$ and expansion of the product lead to required functions φ_p. Fix a function $\alpha : \Re_+ \to [0,1]$ satisfying

$$\alpha(t)=t \ \text{ if } \ |t|\leq 1, \ \alpha(t) = 0 \ \text{ if } \ |t|>2 \ \text{ and } \ |\alpha'|\leq 2. \tag{11}$$

Consider a map $f : \mathcal{E} \to Y$. By translation $f+a$ $(a \in Y)$, we may ensure that f ranges in the ball with center o and radius $2\|f\|_{lip}$. Define the function $\tilde{f} : E \to Y$ as

$$\tilde{f}(x) = \Sigma_p\, f(x_p)\varphi_p\left(\frac{x}{\|x\|}\right)\alpha(\|x\|). \tag{12}$$

By (10), (9) it follows that for $x \in \mathcal{E}$

$$\begin{aligned} &|f(x)-\tilde{f}(x)| \leq \Sigma |f(x_p)-f(x)|\varphi_p(x) \leq \\ &\leq \max_{\|x-x_p\|\leq 2\delta} |f(x)-f(x_p)| \leq 2\delta\|f\|_{lip}. \end{aligned} \tag{13}$$

Also by (9), (10), (11) for arbitrary $x, x' \in E$

$$\|\bar f(x)-\bar f(x')\|\le 2\|f\|_{lip}\,|\alpha(\|x\|)-\alpha(\|x'\|)| +$$

$$\|\Sigma_p f(x_p)\varphi_p(\frac{x}{\|x\|})-\Sigma_q f(x_q)\varphi_q(\frac{x'}{\|x'\|})\|\,\alpha(\|x\|)\le$$

$$\le 4\|f\|_{lip}\,\|x-x'\|+\alpha(\|x\|).$$

$$\sup\{\,\|f(x_p)-f(x_q)\|;p,q\in\mathcal{E},\ \|\frac{x}{\|x\|}-x_p\|<2\delta,\ \|\frac{x'}{\|x'\|}-x_q\|<2\delta\}$$

$$\le 4\|f\|_{lip}\,\|x-x'\|+\|f\|_{lip}\,\{4\delta+\|\frac{x}{\|x\|}-\frac{x'}{\|x'\|}\|\ \|x\|\}$$

and thus

$$\|\bar f(x)-\bar f(x')\|\le 6\|f\|_{lip}\,(\|x-x'\|+\delta). \tag{14}$$

Notice that this estimate does not depend on the smoothness of φ_p. for $0<\tau<1$, define the linear operator u_τ by

$$u_\tau f=\bar f * K_\tau. \tag{15}$$

By (6), (14)

$$\|u_\tau f(x)-u_\tau f(x')\|\le 6\|f\|_{lip}\,(\|x-x'\|+\delta) \tag{16}$$

while by (6),(7), (14)

$$\|\bar f(x)-u_\tau f(x)\|\le\int\|\bar f(x)-\bar f(x+y)\|\,K_\tau(y)\,dy\le 6(\tau+\delta)\|f\|_{lip}$$

and in particular for $x\in\mathcal{E}$, by (13)

$$\|f(x)-u_\tau f(x)\|\le 8(\tau+\delta)\|f\|_{lip}. \tag{17}$$

Our next purpose is to verify estimate (3) for the map $u_\tau f$. This will complete the proof. At this point, we will make use of the special construction of K. The following identity will be exploited.

LEMMA 4: If a,b are compactly supported functions in $L^\infty(\Re^n)$, then

$$(a*b)_{kx}-(a*b)-k[(a*b)_x-(a*b)]=\sum_{0\le j<k}(a_x-a)*(b_{jx}-b) \tag{18}$$

for $k=1,2,\ldots$ and where $a_x(y)=a(y+x)$ is the translate ($x\in\Re^n$).

The proof of (18) is straightforward verification and left to the reader. Apply (18) with $a=\chi_{\tau/2}$, $b=K_{\tau/2}$. Notice that then $a*b=K_\tau$ by (8). Therefore

$$(K_\tau)_{kx}-K_\tau-k[(K_\tau)_x-K_\tau]=\sum_{0\le j<k}[(\chi_{\tau/2})_x-\chi_{\tau/2}]*[(K_{\tau/2})_{jx}-K_{\tau/2}]$$

and convolution with $\bar f$ gives the identity

$$(u_\tau f)_{kx}-u_\tau f-k[(u_\tau f)_x-u_\tau f]$$
$$=\sum_{0\le j<k}[(\chi_{\tau/2})_x-\chi_{\tau/2}]*[(u_{\tau/2}f)_{jx}-u_{\tau/2}f]. \tag{19}$$

Thus by (16), (19) we have pointwise inequality

$$k\,\|(u_\tau f)_x - u_\tau f\| \le 6\|f\|_{lip}(k\|x\|+\delta)$$
$$+\sum_{0\le j<k}\|u_{\tau/2}f\|_{lip}\, j\,\|x\|\;\|\chi_{\tau/2}-(\chi_{\tau/2})_x\|_1 .$$

Remembering the definition of $\chi=\chi_B$ (B convex symmetric, $Vol_n=1$)

$$\|\chi_{\tau/2}-(\chi_{\tau/2})_x\|_1=\|\chi-\chi_{2x/\tau}\|_1\le cn\tau^{-1}\|x\|.$$

So

$$\|(u_\tau f)_x-u_\tau f\|\le 6\|f\|_{lip}(\|x\|+\frac{\delta}{k})+Ckn\tau^{-1}\|u_{\tau/2}f\|_{lip}\;\|x\|^2 \tag{20}$$

for $x\in E$. Fix $0<\varepsilon<1$ (to be precised later) and let $x\in E$. If $\|x\|>\delta\varepsilon$, then (16) implies

$$\|(u_\tau f)_x-u_\tau f\|\le 6(1+\frac{1}{\varepsilon})\|f\|_{lip}\,\|x\|. \tag{21}$$

If $\|x\|<\delta\varepsilon$, choose k satisfying $k\|x\|\sim\delta\varepsilon$ and apply (20)

$$\|(u_\tau f)_x-u_\tau f\|\le 6(1+\frac{1}{\varepsilon})\|f\|_{lip}\,\|x\| + C\delta\tau^{-1}n\varepsilon\|u_{\tau/2}f\|_{lip}\,\|x\|. \tag{22}$$

Taking ε appropriately, $\varepsilon\sim\min(1,(\delta\tau^{-1}n)^{-1})$, it results from (21), (22) that

$$\begin{aligned}&\|(u_\tau f)_x-u_\tau f\|\le c(1+\delta\tau^{-1}n)\|f\|_{lip}\,\|x\|+\frac{1}{10}\|u_{\tau/2}f\|_{lip}\,\|x\|\\ &\|u_\tau f\|_{lip}\le C(1+\delta\tau^{-1}n)\|f\|_{lip}+\frac{1}{10}\|u_{\tau/2}f\|_{lip}.\end{aligned} \tag{23}$$

Repeating (23) with τ replaced by $\frac{\tau}{2}$ yields

$$\|u_\tau f\|_{lip}\le C(1+\delta\tau^{-1}n)(1+\frac{1}{5})\|f\|_{lip}+\frac{1}{100}\|u_{\tau/4}f\|_{lip}$$

and iterating

$$\|u_\tau f\|_{lip}\le C(1+\delta\tau^{-1}n)1+\frac{1}{5}+\cdots+\frac{1}{5^{j-1}})\|f\|_{lip}+10^{-j}\|u_{2^{-j}\tau}f\|_{lip}. \tag{24}$$

Notice that

$$\|u_{2^{-j}\tau}f\|_{lip}\le\|\tilde f\|_{lip}<\infty$$

since the functions φ_p,α were taken smooth. Therefore (24) implies (3).

3. PROOF OF THEOREM 2

Let $\mathcal{E}$ be a $\frac{1}{n}$-net in the Euclidean sphere and $f:\mathcal{E}\to Y$ a one-to-one map in a Banach space Y. Fix $\tau\sim\frac{1}{dist\ f}$, $dist\ f=\|f\|_{lip}\,\|f^{-1}\|_{lip}$ and apply Theorem 1 to obtain $\bar f$. Thus $F=\bar f\,|_{S_{n-1}}$ satisfies

$$\|F\|_{lip}\le c\ dist\ f\cdot\|f\|_{lip}\quad\text{(by (3))}$$

$$\|f(x)-F(x)\|\le\tau\|f\|_{lip}\quad\text{(by (2))}.$$

Let $x, x' \in S_{n-1}$, $|x-x'| > \frac{3}{4}$ and choose $x_0, x'_0 \in \mathcal{E}$ with

$$|x-x_0| < \frac{1}{n} \quad \text{and} \quad |x'-x'_0| < \frac{1}{10}.$$

Clearly

$$\begin{aligned} \|F(x)-F(x')\| &\geq \|F(x_0)-F(x'_0)\| - \frac{2}{n}\|F\|_{lip} \\ &\geq \|f(x_0)-f(x'_0)\| - 2\tau\|f\|_{lip} - \frac{2}{n}\|F\|_{lip} \\ &\geq \frac{1}{2\|f^{-1}\|_{lip}} - 2\tau\|f\|_{lip} - \frac{c}{n}\tau^{-1}\|f\|_{lip} \\ &\geq \frac{1}{3\|f^{-1}\|_{lip}} \end{aligned} \tag{25}$$

for a choice $\tau \sim (dist\ f)^{-1}$ mentioned above and provided $dist\ f \ll n^{1/2}$, which I clearly may assume. In order to linearize f, I use the same simple trick as exploited by Maurey and Pisier in a simplification of Dvoretzky's theorem on spherical sections. It results from (25)

$$\begin{aligned} \frac{c}{\|f^{-1}\|_{lip}} &\leq \int_{S\times S} \|F(x)-F(y)\|\, dxdy = \int_{S\times S} \Big\| \int_0^{\pi/2} \frac{d}{d\theta} F(x\cos\theta + y\sin\theta)\, d\theta \Big\|\, dxdy \\ &\leq \int_{S\times S\times[0,\frac{\pi}{2}]} \|DF(x\cos\theta+y\sin\theta)(-x\sin\theta+y\cos\theta)\|\, dxdy\, d\theta \\ &= \int_{S\times S} \|DF(x')(y')\|\, dx'dy'. \end{aligned} \tag{26}$$

Here dx refers to the normalized invariant measure on $S = S_{n-1}$ and we use the fact that $(x,y) \to (x\cos\theta + y\sin\theta, -x\sin\theta + y\cos\theta)$ is measure preserving. By (26), for some $x' \in S$, the linear map $T : \mathfrak{R}^n, |\ | \to Y$ will satisfy

$$\|T\| \leq \|F\|_{lip} \leq c\,(dist\ f)\|f\|_{lip} \tag{27}$$

$$\int_S \|Tx\|\, dx \geq \frac{c}{\|f^{-1}\|_{lip}}. \tag{28}$$

The conclusion then immediately follows from the results in [FLM].

4. APPLICATION TO NON-LINEAR COTYPE

In [BMW], a notion of "non-linear" type for general metric spaces was developed, generalizing type in normed spaces (see [Mil-Sch] for the linear theory). It turns out to be more difficult to state a simple formulation of non-linear cotype. The purpose of this section is to describe the consequences of Lipschitz embedding of a discrete net $\mathcal{E}$ of points in the Cartesian cube $[0,1]^n$ is a given Banach space Y.

PROPOSITION 5: Let $f : \mathcal{E} \to Y$ be a one-to-one map where $\mathcal{E}$ is a $\frac{1}{n}$-net in

$[0,1]^n$ considered as subset of ℓ_n^∞. Then Y contains ℓ_n^∞-subspaces, $m \geq \delta(dist\ f)\cdot n$, up to an isomorphism factor $\sim (dist\ f)^2$.

Proof: Application of Theorem 1 yields a map $F : [0,1]^n \to Y$ satisfying

$$\|f(x)-F(x)\| \leq \tau \|f\|_{lip} \quad \text{for} \quad x \in \mathcal{E}$$

$$\|F\|_{lip} \leq c(1+\tau^{-1}\|f\|_{lip}).$$

Clearly for $\varepsilon_i = \pm 1$

$$\|F\|_{lip} \geq \| \frac{d}{dt} F(x_1+t\varepsilon_1, \ldots, \alpha_n+t\varepsilon_n)|_{t=0}\| = \|\Sigma \partial_i F(x)\cdot\varepsilon_i\|. \tag{29}$$

A similar computation as in the previous section shows that for $\tau \sim (dist\ f)^{-1}$

$$\frac{1}{3\|f^{-1}\|_{lip}} \leq \|F(x_1, \ldots, x_{j-1}, 1, x_{j+1}, \ldots, x_n) - F(x_1, \ldots, x_{j-1}, 0, x_{j+1}, \ldots, x_n)\|$$

$$\leq \int_0^1 \|\partial_j F(x)\| dx_j.$$

Hence for $j = 1, \ldots, n$

$$\int_{[0,1]^n} \|\partial_j F(x)\| dx \geq \frac{1}{3\|f^{-1}\|_{lip}}.$$

Choose x such that

$$\sum_{j=1}^n \|\partial_j F(x)\| \geq \frac{n}{3\|f^{-1}\|_{lip}}. \tag{30}$$

Then by (29), (30), $\xi_j = \partial_j F(x)$ satisfy

$$\|\Sigma \varepsilon_j \xi_j\| \leq \|F\|_{lip} \quad (\varepsilon_j = \pm 1)$$

$$\Sigma \|\xi_j\| \geq \frac{n}{3\|f^{-1}\|_{lip}}.$$

The ℓ_m^∞-subspaces are then obtained by well-known and simple arguments.

Denote a usual by $c_q(Y)$ the cotype q constant of the Banach space Y. $(2 \leq q < \infty)$.

PROPOSITION 6: Let $f : \{1,2, \ldots, [n^{1+1/q}]\}^n \to Y$ be a one-to-one map. Then

$$dist\ f \geq \frac{c}{c_q(Y)} n^{1/q}. \tag{31}$$

Here the metric on $\{1,2, \ldots, N\}^n$ is the maximum coordinate difference.

Proof: Assume $dist\ f < n^{1/q}$. Again using Theorem 1 with $E = \ell_n^\infty$, $\mathcal{E} = n^{-1-1/q}$ net in unit cube, $\tau \sim (dist\ f)^{-1}$, a differentiable map $F : [0,1]^n \to Y$ is obtained satisfying

$$C(dist\ f)\|f\|_{lip} n^{-1/q} \geq \|F\|_{lip} \geq \|\sum_{j=1}^n \partial_j F(x)\varepsilon_j\| \quad (\varepsilon_j = \pm 1) \tag{32}$$

$$\int_{[0,1]^n} \| \partial_j F(x) \| \, dx \geq \frac{1}{3 \| f^{-1} \|_{lip}} \quad (cf. (25)). \tag{33}$$

Hence, by integration

$$C \| f \|_{lip} \geq \iint \| \sum_{j=1}^{n} \partial_j F(x) \varepsilon_j \| \, d\varepsilon dx \geq \frac{1}{c_q(Y)} \int (\Sigma \| \partial_j F(x) \|^q)^{1/q}$$

$$\geq \frac{1}{3C_q(Y)} \frac{n^{1/q}}{\| f^{-1} \|_{lip}}.$$

5. A PROOF OF RIBE'S THEOREM

Assume E an n-dimensional normed space and $\mathcal{E}$ a δ-net in the unit ball. Let us point out that δ will be here of a different order of magnitude than $\frac{1}{n}$, since we will need $n \sim \log \log \frac{1}{\delta}$. Let $f : \mathcal{E} \to Y$ be a Lipschitz map (meaning that we bound $dist\ f$). Application of Theorem 1 yields a smooth map $F : E \to Y$ fulfilling the conditions

$$\| f(x) - F(x) \| < \delta' \quad \text{for } x \in \mathcal{E} \tag{34}$$

$$\| F \|_{lip} \leq C \| f \|_{lip}. \tag{35}$$

Roughly speaking, δ' in (34) is "arbitrarily small" and C in (35) is an absolute constant. Notice that a much cruder dependence on the dimension n in (3) would suffice. Also, with some additional care, the constant C may be taken any fixed number > 1. Denote $(P_t)_{t>0}$ the kernel of the Poisson-semigroup on $\Re^n$. Our purpose is to show that for suitable t the function

$$F * P_t$$

will satisfy

$$\| \partial_a (F * P_t)(x) \| \geq \frac{1}{2 \| f^{-1} \|_{lip}} \tag{36}$$

for some point $x \in E$ and a taken in κ-net in the unit sphere S of E. Considering then $T = D(F * P_t)(x)$ a linear operator from E to Y is obtained for which $\| T \| \leq \| F \|_{lip} \leq C \| f \|_{lip}$ and for a in this κ-set $\| Ta \| \geq \frac{1}{2} \| f^{-1} \|_{lip}^{-1}$. Hence for arbitrary $x \in S$

$$\| Tx \| \geq \frac{1}{2} \| f^{-1} \|_{lip}^{-1} - \kappa \| T \| \geq \frac{1}{3} \| f^{-1} \|_{lip}^{-1}$$

for κ small enough. Thus T is the desired embedding of E into Y. It remains to achieve (36). Fix $a \in S$ and notice that $\| \partial_a (F * P_t) \|$ is subharmonic, in the sense that

$$\| \partial_a (F * P_t) \| * P_s \geq \| \partial_a (F * P_{t+s}) \|. \tag{37}$$

As a consequence of (34), for $x, x' \in \frac{B}{2}$.

$$\|F(x)-F(x')\| \geq \frac{9}{10}\|f^{-1}\|_{lip}^{-1}\|x-x'\| \tag{38}$$

$$\|(F * P_t)(x)-(F * P_t)(x')\| \geq \frac{9}{10}\|f^{-1}\|_{lip}^{-1}\|x-x'\|. \tag{39}$$

(38) under the restriction

$$\|x-x'\| \geq M\delta$$

(39) under the restriction

$$\|x-x'\| \geq \sqrt{n}\ Mt$$

where M depends on $\|f\|_{lip}$ and t is taken large enough w.r.t. δ. We assume here that

$$\frac{1}{\sqrt{n}}|x| \leq \|x\| \leq |x| \tag{40}$$

which is obtained letting the maximal volume ellipsoid of E coincide with the Euclidean ball.

From (39) we get for $x \in \frac{1}{2}B$ and $a \in S$, $M_1 = \sqrt{n}\ M$

$$\begin{aligned} &\frac{1}{M_1 t}\|(F * P_t)(x+M_1 ta)-(F * P_t)(x)\| \geq \frac{4}{5}\|f^{-1}\|_{lip}^{-1} \\ &\frac{1}{M_1 t}\int_0^{M_1 t}\|\partial_a(F * P_t)(x+sa)\|ds \geq \frac{4}{5}\|f^{-1}\|_{lip}^{-1}. \end{aligned} \tag{41}$$

Let R be a number to be defined later, depending on n and *dist* f. From (41), we derive for t sufficiently small, on $\frac{B}{3}$

$$\begin{aligned} &\frac{1}{M_1 t}\int_0^{M_1 t}\{\|\partial_a(F*P_t)\| * (P_{Rt})_{sa}\{ds = \\ &\frac{1}{M_1 t}\int_0^{M_1 t}\int_{\Re^n}\|\partial_a(F*P_t)(x-y+sa)\|P_{Rt}(y)dy\ ds \geq \frac{7}{10}\|f^{-1}\|_{lip}^{-1}. \end{aligned} \tag{42}$$

It follows from (40) that

$$\|(P_{Rt})_{sa}-P_{Rt}\|_1 = \|P_1-(P_1)_{\frac{sa}{Rt}}\|_1 \leq C\sqrt{n}\frac{s|a|}{Rt} \leq C\ n\ \frac{s}{Rt} \tag{43}$$

and using (42), on $\frac{B}{3}$

$$\|\partial_a(F*P_t)\| * P_{Rt} \geq \frac{7}{10}\|f^{-1}\|_{lip}^{-1} - C\|f\|_{lip}\frac{M_1}{R}n > \frac{6}{10}\|f^{-1}\|_{lip}^{-1} \tag{44}$$

for appropriate choice of R.

By (37), we have pointwise inequality

$$\|\partial_a(F*P_t)\| * P_{Rt} \geq \|\partial_a(F * P_{(R+1)t})\|. \tag{45}$$

Now a simple reasoning shows that taking a in a κ-net $\mathcal{F}$, we may obtain for some $t > 0$ (sufficiently large w.r.t δ)

$$\int_{\Re^n} [\|\partial_a(F*P_t)\| * P_{Rt}] = \int_{\Re^n} \|\partial_a(F*P_t)\| \approx \int_{\Re^n} \|\partial_a(F*P_{(R+1)t})\|$$

which means, that, by (45)

$$\|\partial_a(F*P_t)\| * P_{Rt} \approx \|\partial_a(F * P_{(R+1)t})\|$$

and hence, by (44)

$$\|\partial_a(F*P_{(R+1)t})\| \geq \frac{1}{2}\|f^{-1}\|_{lip}^{-1} \tag{46}$$

"almost everywhere" on the set $B/3$. Thus to obtain (36), it suffices to take $x \in \frac{B}{3}$ where (46) is fulfilled for each $a \in \mathcal{F}$. This is possible, since for each $a \in \mathcal{F}$ the "exceptional set" has small measure.

A more quantitative analysis leads to the condition $\log\log \sim n \log(dist\ f)$. The method described above also permits to obtain the almost isometric result. Thus if K appearing in (1) tends to 1, then X is $1+\varepsilon(K)$-finitely representable in Y.

REMARK: In the case Y is a uniformly convex space the previous reasoning simplifies considerably since it can be shown that f itself is "locally linear". This fact was pointed out to me by J. Lindenstrauss.

6. PROOF OF THEOREM 3

We show that if $F : \ell^2 \to \ell^p$ $(p < 2)$, then F does not satisfy (1). The particular case $p = 1$ is due to P. Enflo and what follows is a modification of his argument.

Consider the decreasing sequence

$$A_n = \sup_{\|x-y\| \geq n} \frac{\|F(x)-F(y)\|}{\|x-y\|}$$

and $A = \lim_{n\to\infty} A_n$. Fix $\tau > 0$ (to be defined later) and n_0 satisfying

$$A_{n_0} \leq A + \tau. \tag{50}$$

Take $x, y \in \ell^2$ such that

$$2d \equiv \|x-y\| > 2n_0 \tag{51}$$

$$\|F(x)-F(y)\| > (A_{2n_0}-\tau)\|x-y\|. \tag{52}$$

Obviously, we may assume x, y supported by a finite number of coordinates. For e a unit vector outside the support of x, y, define

$$z_e = \frac{x+y}{2} + \delta d e$$

where $0 < \delta < 1$ will be fixed later. Clearly

$$\|z_e - x\| = \|z_e - y\| = (1+\delta^2)^{1/2} d > n_0 \tag{53}$$

and hence by definition of A_{n_0} and (50)

$$\begin{aligned} \|F(x) - F(z_e)\| &\leq (A+\tau)(1+\delta^2)^{1/2} d \\ \|F(y) - F(z_e)\| &\leq (A+\tau)(1+\delta^2)^{1/2} d \end{aligned} \tag{54}$$

Let I be a finite set such that (up to a negligible error) $F(x), F(y)$ belong to $[e_i \mid i \in I]$. Denote π_I the coordinate projection. By triangle inequality

$$\|F(x) - \pi_I F(z_e)\| + \|F(y) - \pi_I F(z_e)\| \geq \|F(x) - F(y)\| > 2(A-\tau)d. \tag{55}$$

Also

$$\begin{aligned} &\|F(x) - F(z_e)\|^p = \|\pi_I F(x) - \pi_I F(z_e)\|^p + \|\pi_{I^c} F(x) - \pi_{I^c} F(z_e)\|^p \geq \\ &\|F(x) - \pi_I F(z_e)\|^p + \|\pi_{I^c} F(z_e)\|^p - \tau \end{aligned} \tag{56}$$

and similarly replacing x by y. (57).

By (54), (56), (57), (55)

$$\begin{aligned} &2(A+\tau)^p (1+\delta^2)^{p/2} d^p \geq \\ &\|F(x) - \pi_I F(z_e)\|^p + \|F(y) - \pi_I F(z_e)\|^p + 2\|\pi_{I^c} F(z_e)\|^p - 2\tau \geq \\ &2^{-p/p'}[2(A-\tau)d]^p + 2\|\pi_{I^c} F(z_e)\|^p - 2\tau. \end{aligned} \tag{58}$$

Notice that for $e \neq e'$, $\|z_e - z_{e'}\| = 2\delta d > 1$ for d large enough. Hence, by (1)

$$\|F(z_e) - F(z_{e'})\| \geq K^{-1} \|z_e - z_{e'}\| = 2\delta d K^{-1}.$$

Since I is finite, a simple compactness argument will yield unit vectors e for which

$$\|\pi_{I^c} F(z_e)\| \geq \frac{1}{2} \delta d K^{-1}. \tag{59}$$

Substitution of (59) in (58) yields

$$(A+\tau)^p (1+\delta^2)^{p/2} - (A-\tau)^p \geq (\frac{1}{2} K^{-1} \delta)^p - \tau.$$

This holds for all $\tau > 0$, so

$$A^p (1+\delta^2)^{p/2} - A^p \geq (\frac{1}{2} K^{-1} \delta)^p$$

and a contradiction is found letting $\delta \to 0$, since $p < 2$. the proof is completed.

References

[Ben] Y. BENJAMINI: The uniform classification of Banach spaces (expository paper), Longhorn notes, University of Texas, Austin, 1984-85, 15-39.

[BMW] J. BOURGAIN, V. D. MILMAN, H. WOLFSON: On type of metric spaces, Trans. Amer. Math. Soc. 294 (1986), 205-317.

[FLM] T. FIGIEL, J. LINDENSTRAUSS, V. D. MILMAN: The dimension of almost spherical sections of convex bodies, Acta Math. 139 (1977), 53-94.

[Mil, Sch.] V. D. MILMAN, G. SCHECHTMAN, Asymptotic theory of finite dimensional normed spaces, Springer LNM, Vol 1200.

ON DIMENSION FREE MAXIMAL INEQUALITIES FOR CONVEX SYMMETRIC BODIES IN $\mathfrak{R}^n$

J. BOURGAIN

IHES, France and University of Illinois

1. INTRODUCTION

The purpose of this note is to elaborate on Remark (5) at the end of [B1] and show how, in certain cases, the method described in [B1] enables us to prove dimension free bounds on M_B for $p \leq \frac{3}{2}$. Here

$$M_B f(x) = \sup_{t>0} \frac{1}{Vol\ B} \int_B |f(x+ty)|\, dy$$

is the maximal operator associated to the convex symmetric body B in $\mathfrak{R}^n$. It was proved in [B1] that

$$\| M_B f \|_{L^p(\mathfrak{R}^n)} \leq c(p) \| f \|_{L^p(\mathfrak{R}^n)}$$

or for short

$$\| M_B \|_{p \to p} \leq c(p)$$

if $p > \frac{3}{2}$ and where $c(p)$ is an absolute constant only dependent on p. In the case of the euclidean ball, the restriction $p > \frac{3}{2}$ may be replaced by $p > 1$. The result was proved by E. Stein and T. O. Stromberg in [S-S], using different arguments. Their method is based on the euclidean symmetry and hence only applies to this specific class. My purpose here is to give an alternative approach, only using the properties of the Fourier-transform

$$\hat{\chi}_B(\xi) = \int_B e^{-2\pi i \langle x, \xi \rangle} dx$$

of the indicator function χ_B of B and no additional geometric considerations. This method enables us to extend the euclidean result to the classes

$$B_s^{(n)} = \{x \in \mathfrak{R}^n \mid \sum_{j=1}^n x_j^{2s} \leq 1\}$$

for $s = 1,2,\ldots$

The main point in the argument is to obtain a dimension free estimate on maximal functions

$$\sup_{t>0} |f * K_t| \,;\, K_t(x) = t^{-n} K(t^{-1}x)$$

where $K \geq 0$, $K \in L^1(\mathfrak{R}^n)$ only involving the Fourier transform $\hat{K}$. Such a result is stated and proved in the next section of this paper, repeating part of [B1]. Results along this line were obtained independently by T. Carbery.

2. A GENERAL ESTIMATE

As usual, $(P_t)_{t>0}$ stands for the Poisson semigroup on $\mathfrak{R}^n$, $\hat{P}_t(\xi) = e^{-t|\xi|}$ where $|\xi|$ is the euclidean norm. Recall the maximal and g-function inequalities (see [S], section 2) for $1 < p < \infty$, $u(x,t) = (f * P_t)(x)$

$$\| \sup_{t>0} |f * P_t| \|_p \le C(p) \|f\|_p \tag{2.1}$$

$$\| \{ \int_0^\infty | \frac{\partial u}{\partial t} |^2 t dt \}^{1/2} \|_p \le \| \{ \int_0^\infty |\nabla u(x,t)|^2 t dt \}^{1/2} \|_p \le c(p) \|f\|_p \tag{2.2}$$

where $\| \ \|_p$ denotes the $L^p(\mathfrak{R}^n, dx)$ norm. The feature of these inequalities which we like to emphasize here is the fact that the constant $C(p)$ does not depend on the dimension n of the space.

PROPOSITION: Let $K \in L^1(\mathfrak{R}^n)$, $K \ge 0$ and $\hat{K}$ satisfying

$$|\hat{K}(\xi)| < C |\xi|^{-c} \ ; \ |1 - \hat{K}(\xi)| < C |\xi|^c \tag{2.3}$$

for some constants c, C. Then for $1 < p < \infty$

$$\| \sup_{t \in I_1} |f * K_t| \|_p \le A(c, C, p) \|f\|_p \tag{2.4}$$

where

$$I_1 = \{ 2^j \mid j \in Z \}.$$

Let further η be a positive integer such that

$$|\partial_t^{(j)} \hat{K}(t\xi)|_{t=1}| < C \quad \text{for } j \le \eta \tag{2.5}$$

Then

$$\| \sup_{t>0} |f * K_t| \|_p \le A'(c, C, p) \|f\|_p \quad \text{for } p > 1 + \frac{1}{2\eta} \tag{2.6}$$

The inequality (2.6) is essentially optimal as one verifies by considering the spherical maximal functions. Again the point here is to get dimension independent bounds. Inequality (2.4) holds if $p = 2$ and hence for $2 \le p \le \infty$, since replacing K by $\tilde{K} = K - P_1$ it follows from Parseval's identity

$$\| \sup_{t \in I_1} |f * \hat{\tilde{K}}_t| \|_2^2 \le \sum_{j \in Z} \int |\hat{f}(\xi)|^2 |\hat{\tilde{K}}(2^j \xi)|^2 d\xi < C \|f\|_2^2 .$$

since, by (2.3)

$$|\hat{K}(2^j\xi)| \le C \min(2^j|\xi|, \frac{1}{2^j|\xi|})^c .$$

Thus the $\hat{K}(2^j\xi)$ are almost disjointly supported.

Consider now the case $1 < p \le 2$. Following [B1], denote for $1 \le q \le \infty$ by $A(p,q)$ the best constant in the inequality

$$\|(\sum_{j \in Z} |f_j * K_{2^j}|^q)^{1/q}\|_p \le A(p,q) \|(\Sigma |f_j|^q)^{1/q}\|_p$$

Thus (2.4) amounts to finding uniform bounds on $A(p,\infty)$, $p > 1$ if K fulfils (2.3). Notice that

$$A(p,1) = A(p',\infty) \quad \text{(duality)} \tag{2.7}$$

$$A(p,2) \le A(p,1)^{1/2} A(p,\infty)^{1/2} \quad \text{(interpolation)} \tag{2.8}$$

(2.7), (2.8) and the estimate on $A(p',\infty) \le A(2,\infty)$ permit thus to write

$$A(p,2) \le C\ A(p,\infty)^{1/2} \tag{2.9}$$

We now prove a "converse" to (2.9) estimating $A(q,\infty)$ in terms of $A(p,2)$ for $p < q$. Replace again K by $\tilde{K}$ and write

$$\|\sup_j |f*(\tilde{K})_{2^j}|\ \|_q \le \|(\Sigma_j |f*(\tilde{K})_{2^j}|^2)^{1/2}\|_q .$$

Using the g-function inequality (2.2) and interpolation arguments, it follows for $q > p > 1$

$$A(q,\infty) \le C.c(p).(q-p)^{-2} A(p,2) \tag{2.10}$$

The reader is referred to [B1] for the details. (2.9) and (2.16) give

$$A(q,\infty) \le C.C(p).(q-p)^{-2} A(p,\infty)^{1/2} \tag{2.11}$$

In the previous discussion, we assumed the $A(p,\infty)$ a priori finite. This can be achieved by assuming K compactly supported and replacing K by $K * P_\varepsilon$ where $\varepsilon \to 0$. The final estimate will not depend on this a priori bound. In fact, in the applications the a priori bound is obvious.

Fix $p_0 > 1$ and define

$$T = \sup_{p_0 < p \le 2} A(p,\infty)(p-p_0)^4 \tag{2.12}$$

Take $q > p_0$ such that

$$\frac{T}{2} < A(q,\infty)(q-p_0)^4 \tag{2.13}$$

and $2p = p_0 + q$.

By (2.11), (2.12),(2.13)

$$\frac{T}{2(q-p_0)^4} \le A(q,\infty) \le CC(p)(q-p)^{-2}\frac{T^{1/2}}{(p-p_0)^2} \le 16CC(p)\,\frac{T^{1/2}}{(q-p_0)^4}$$

$$T \le [32C.C(p_0)]^2.$$

Hence $A(p,\infty) \le [32C.C(p_0)]^2(p-p_0)^{-4}$ where $p_0 > 1$ is arbitrary. This proves (2.4).

We now pass to the proof of (2.6). Denote more generally for $r = 1,2,\ldots$

$$I_r = \{2^{j/r} \mid j \in Z\}$$

and

$$M_r f = \sup_{t \in I_r}(f * K_t) \quad (f \ge 0).$$

Thus M_1 is the "dyadic" maximal operator considered in (2.4). Writing

$$I_r = \bigcup_{0 \le \alpha < r} I_r^{(\alpha)}\,;\, I_r^{(\alpha)} = \{2^{j/r}\ ;\, j \in rZ + \alpha\}$$

we get

$$\| M_r \|_{p \to p} \le r \| M_1 \|_{p \to p} \le rA(p) \tag{2.14}$$

Estimate

$$Mf = \lim_{r \to \infty} M_r f \le M_1 f + \sum_{s=0}^{\infty} | M_{2^{s+1}} f - M_{2^s} f | \tag{2.15}$$

One easily verifies that

$$| M_{2r} f - M_r f | \le M'_r f$$

where

$$M'_r f = \sup_{t \in I_r} | f * K(r)_t |$$

$$K(r) = (K_{2^{1/2r}}) - K$$

Hence, if we let $\tilde{M}_s = M'_{2^s}$, it follows from (2.15)

$$Mf \leq M_1 f + \sum_{s=0}^{\infty} \tilde{M}_s f \tag{2.16}$$

Fix s_1 and repeat previous procedure to get

$$\tilde{M}_{s_1} f \leq M_{s_1,1} f + \sum_{0 \leq s \leq s_1} \tilde{M}_{s_1,s} f \tag{2.17}$$

where

$$M_{s_1,1} f = \sup_{t \in I_1} | f * K(2^{s_1})_t | \quad \text{and} \quad \tilde{M}_{s_1,s} f = \sup_{t \in I_{2^s}} | f * K(2^{s_1}, 2^s)_t |$$
$$K(2^{s_1}, r) = K(2^{s_1})_{2^{1/2r}} - K(2^{s_1}) \tag{2.18}$$

Substituting (2.17) in (2.16)

$$Mf \leq M_1 f + \Sigma_{s_1} M_{s_1,1} f + \sum_{s_1 \geq s_2} \tilde{M}_{s_1,s_2} f$$

Iterating η times, one gets

$$M_f \leq M_1 f + \Sigma_{s_1} M_{s_1,1} f + \sum_{s_1 \geq s_2} M_{s_1,s_2,1} f + \cdots + \sum_{s_1 \geq s_2 \cdots \geq s_\eta} \tilde{M}_{s_1,\ldots,s_\eta} f \tag{2.20}$$

For $\nu < \eta$, $M_{s_1,\ldots,s_\nu,1}$ is given by the kernel $K(2^{s_1}, \ldots, 2^{s_\nu})$ and $\tilde{M}_{s_1,\ldots,s_\eta}$ by $K(2^{s_1}, \ldots, 2^{s_\eta})$. These kernels are obtained according to (2.18). Thus for $p > 1$

$$\| M_{s_1,\ldots,s_\nu,1} \|_{p \to p} \leq 2^\nu A(p) \tag{2.21}$$

$$\| M_{s_1,\ldots,s_\eta} \|_{p \to p} \leq 2^\nu 2^{s_\eta} A(p) \tag{2.22}$$

Notice that for $\nu \leq \eta$

$$\begin{aligned} |\hat{K}(2^{s_1}, \ldots, 2^{s_\nu})(\xi)| &\leq 2^{-s_\nu} \sup_{1 \leq t \leq 1+2^{-s_\nu}} |\partial_t \hat{K}(2^{s_1}, \ldots, 2^{s_{\nu-1}})(t\xi)| \\ &\leq 2^{-s_1 - \cdots - s_\nu} \sup_{1 \leq t \leq 2} |\partial_t^{(\nu)} \hat{K}(t\xi)| \\ &\leq C\, 2^{-s_1 - \cdots - s_\nu} \end{aligned} \tag{2.23}$$

where C is the constant appearing in (2.5).

Similarly

$$| \langle \nabla \hat{K}(2^{s_1}, \ldots, 2^{s_\nu})(\xi), \xi \rangle | \leq 2C\, 2^{-s_1 - \cdots - s_{\nu-1}} \tag{2.24}$$

From (2.3), (2.23)

$$|\hat{K}(2^{s_1}, \ldots, 2^{s_\nu})(\xi)| \leq 2^\eta C \min(|\xi|^{-c}, |\xi|^c, 2^{-s_1 - \cdots - s_\nu}) \tag{2.25}$$

At this point, in order to obtain the L^2-bound, invoke Lemma 2 of [B2]

LEMMA: $\| \sup_{t>0} | f * K_t | \|_2 \leq C \Gamma(K) \| f \|_2$

where

$$\Gamma(K) = \sum_{j \in Z} \alpha_j^{1/2} (\alpha_j + \beta_j)^{1/2}$$

and

$$\alpha_j = \sup_{|\xi| \sim 2^j} |\hat{K}(\xi)| \ ; \ \beta_j = \sup_{|\xi| \sim 2^j} |<\nabla \hat{K}(\xi), \xi>|$$

which is applied taking $K = K(2^{s_1}, \ldots, 2^{s_\nu})$.

Clearly by (2.24), (2.25)

$$\Gamma(K(2^{s_1}, \ldots, 2^{s_\nu})) \leq C(\eta)(s_1 + \cdots + s_n) 2^{-s_1 - \cdots - s_{\nu-1} - \frac{s_\nu}{2}} \tag{2.26}$$

Let $p < q < 2$, $\frac{1}{q} = \frac{1-\theta}{p} + \frac{\theta}{2}$. Interpolating and applying (2.21), (2.26) and the lemma

$$\begin{aligned} \| M_{s_1, \ldots, s_\nu, 1} \|_{q \to q} &\leq \| M_{s_1, \ldots, s_\nu, 1} \|_{p \to p}^{1-\theta} \| M_{s_1, \ldots, s_\nu, 1} \|_{2 \to 2}^{\theta} \leq \\ &\leq A(p) C(\eta) 2^{-\frac{\theta}{3}(s_1 + \cdots + s_\nu)} \end{aligned} \tag{2.27}$$

This takes care about the first terms in (2.20). The main contribution is the last term. Now (2.22) and (2.26) with $\nu = \eta$ yield by interpolation

$$\| \tilde{M}_{s_1, \ldots, s_\eta} \|_{q \to q} \leq A(p) C(\eta) 2^{(1-\theta) s_\eta} (s_1 + \cdots + s_\eta)^{\theta} 2^{-\theta[s_1 + \cdots + s_{\eta-1} + \frac{1}{2} s_\eta]}$$

Hence

$$\sum_{s_1 \geq \cdots \geq s_\eta} \| \tilde{M}_{s_1, \ldots, s_\eta} \|_{q \to q} \leq A(p) C(\eta, \theta)$$

provided, as the reader easily verifies

$$\theta > (\eta + \frac{1}{2})^{-1} \tag{2.28}$$

Since we may choose any $p > 1$, (2.28) translates in $q > 1 + \frac{1}{2\eta}$, i.e. (2.6).

This completes the proof of the Proposition.

For general convex symmetric B, after a suitable choice of coordinates, the proposition can always be applied taking for η the value 1. Here $K = \chi_B$. Hence we get absolute bounds on the dyadic maximal operator and on $\| M_B \|_{p \to p}$ as long as $p > \frac{3}{2}$. These are the results of [B1].

3. MAXIMAL-FUNCTION BOUNDS FOR SPECIAL BODIES.

The method described in the previous section yields an alternative proof of the theorem of Stein-Stromberg on the euclidean maximal function. This proof applies moreover to bodies

$$B_s = \{x \in \Re^n \mid \sum_{j=1}^{n} x_j^{2s} \leq 1\}$$

where $s = 1,2,\ldots$ (i.e. ball of ℓ_n^r-spaces for even r).

Fix n. Define m by

$$Vol\ B_s = (\frac{1}{m})^n$$

thus $m \sim n^{1/2s}$ up to a constant and $\tilde{B}_s = mB_s$ is of volume 1. Fix a positive integer η and let $\tau : \Re_+ \to [0,1]$ be a smooth function satisfying the conditions

$$\tau = 1 \quad \text{on} \quad [0,m^{2s}] \tag{3.1}$$

$$\tau = 0 \quad \text{on} \quad [m^{2s}+1,\infty[\tag{3.2}$$

$$|\tau^{(j)}| \leq C(\eta) \quad \text{for} \quad j \leq \eta \tag{3.3}$$

Thus by (3.1)

$$\chi_{\tilde{B}_s}(x) \leq \tau(\sum_{j=1}^{n} x_j^{2s}) = K(x)$$

and therefore

$$M_{B_s} f \leq \sup_{t>0} | f * K_t |.$$

On the other hand, K is supported by the set

$$(m^{2s}+1)^{1/2s} B_s \subset (1+\frac{C}{n})\tilde{B}_s \tag{3.4}$$

and thus $\| K \|_{L^1(\Re^n)} \leq C$.

In order to apply the Proposition, let us verify condition (2.5). We get by partial integration

$$|\partial_t^{(j)}\hat{K}(t\xi)|_{t=1} \sim \int e^{2\pi i<x,\xi>}<x,\xi>^j K(x)dx \sim \int e^{2\pi i<x,\xi>}\partial_\zeta^{(j)}[K(x)<x,\zeta>^j]dx \tag{3.5}$$

where $\zeta = \dfrac{\xi}{|\xi|}$ and ∂_ζ is the derivation in direction ζ. Estimate (3.5) by

$$\int |\partial_\zeta^{(j)}[K(x)<x,\zeta>^j]|dx \tag{3.6}$$

giving rise to expressions of the form ($k \le \eta$)

$$\int |\partial_\zeta^{(k)}K(x)| \, |<x,\zeta>|^k dx \tag{3.7}$$

Notice that K and its derivatives are supported by $(1+\dfrac{C}{n})\tilde{B}_s$, from (3.4). Estimate (3.7) by Hölder's inequality

$$C \, \|\partial_\zeta^{(k)}K(x)\|_{L^2(\tilde{B}_s)} . \|<x,\zeta>^k\|_{L^2(\tilde{B}_s)}$$

The second factor can be estimated using the equivalence (see [B2])

$$\|<x,\xi>\|_{L^2(\tilde{B}_s)} \sim \|<x,\xi>\|_{L^{\psi_1}(\tilde{B}_s)} \tag{*}$$

(weak generalization of Khintchine's inequality for convex symmetric bodies of volume 1).

Hence (3.7) is bounded by

$$\|\partial_\zeta^{(k)}K(x)\|_{L^2(\tilde{B}_s)} \tag{3.8}$$

up to a constant depending on η.

Using the definition of K and the chain rule, this expression is easy to evaluate. Let us handle the case $k = 2$ for instance. Thus

$$\partial_\zeta K = 2s \, \tau'(\Sigma x_j^{2s}) \, \Sigma \, x_j^{2s-1}\zeta_j$$

and

$$\partial_\zeta^2 K = 4s^2\tau''(\Sigma x_j^{2s})(\Sigma x_j^{2s-1}\zeta_j)^2 + 2s(2s-1)\tau'(\Sigma x_j^{2s})(\Sigma x_j^{2s-2}\zeta_j^2)$$

from where

$$\|\partial_\zeta^2 K\|_2^2 \le C(s)\int_{\tilde{B}_s} |\Sigma x_j^{2s-1}\zeta_j|^4 dx + C(s)\int_{\tilde{B}_s} |\Sigma x_j^{2(s-1)}\zeta_j^2|^2 dx \tag{3.9}$$

Since the map $x \to \varepsilon x$, $\varepsilon \in \{1,-1\}^n$ is a measure preserving transformation of $\tilde{B}_s$, (3.9) is finally estimated up to $C(s)$ by

$$\sup_j \| x_j \|_{L^{\psi_1}(\tilde{B}_s)} \leq const.$$

There are some straight computations here which we leave to the reader. The last inequality follows from (*). Notice that (*) is valid beyond the class of ℓ_n^s-unit balls. Actually (*) follows from the Brunn-Minkowski inequality valid for a general convex symmetric body. Giving an explicit proof for $\tilde{B}_s$ would be more complicated than proving it by the general geometric considerations in [B2].

Thus finally, by the proposition, one gets

THEOREM: $\| M_{B_s} \|_{p \to p} \leq C(p,s)$ independent of the dimension, for all $p > 1$.

REFERENCES

[B1] J. Bourgain: On the L^p-bounds for maximal functions associated to convex bodies in $\Re^n$, Israel J. Math., 54 (1986), 257-265.

[B2] J. Bourgain: On high dimensional maximal functions associated to convex bodies, Amer. J. Math., Vol 108, N6, 1986, 1467-1476.

[S] E.M. Stein: Topics in Harmonic Analysis, Ann. Math. Studies No. 63, Princeton UP, 1970.

[S-S]
E. M. Stein, J. O. Stromberg: Behavior of maximal functions in $\Re^n$ for large n, Arkiv Math. 21 (1983), 259-269.

ON LIPSCHITZ EMBEDDING OF FINITE METRIC SPACES IN LOW DIMENSIONAL NORMED SPACES

William B. JOHNSON[1,2]
Texas A&M University
College Station, Texas.

Joram LINDENSTRAUSS[1]
The Hebrew University
Jerusalem.

Gideon SCHECHTMAN[1]
The Weizmann Institute of Science
Rehovot.

Given two metric spaces (M,d) and (M',d'), we say that (M,d) C-embeds into (M',d') if there exists a function $f : M \underset{onto}{\overset{1-1}{\to}} N \subset M'$ such that $\|f\|_{Lip}\|f^{-1}\|_{Lip} \le C$ where

$$\|f\|_{Lip} = \sup_{\substack{s\neq t \\ in\ M}} \frac{d'(f(s),f(t))}{d(s,t)} .$$

We report here on some progress on the following problem ([JL] and [B]) : Given $1 \le C < \infty$ and $m \in \mathbb{N}$, what is the smallest k such that any metric space with m points C-embeds into a normed space of dimension k.

THEOREM: *For some absolute constant K and any $0 < \beta \le 1$, any finite metric space with m points $\frac{K}{\beta}$-embeds into a normed space of dimension k where*

$$k \le \begin{cases} Km^\beta & \text{if} \quad \dfrac{10\log_2(\log_2 m)}{\log_2 m} \le \beta \\ K\beta^{-2}(\log_2 m)^3 m^\beta & \text{if} \quad \dfrac{1}{\log_2 m} \le \beta < \dfrac{10\log_2(\log_2 m)}{\log_2 m} \\ K\log_2 m & \text{if} \quad \beta < \dfrac{1}{\log_2 m} \end{cases}$$

(The last case was proved in [B]).

---

[1] *Supported in part by U.S.-Israel B.S.F.*

[2] *Supported in part by N.S.F.*

Most of this paper is devoted to the proof of the theorem. Let us describe briefly the idea of the proof. Given a finite metric space (M, d) we first 1-embed it into some specific normed space (we shall actually use a very specific embedding into some ℓ_∞^n space). We then consider the set $R = \{\frac{x-y}{d(x,y)};\ x \neq y,\ x, y \in M\}$. R is a symmetric set and each $r \in R$ is on the boundary of $convR$. We then find a linear operator T with small rank which approximately preserves this property, i.e., for each $r \in R$, $\frac{K}{\alpha}Tr$ will be outside the set $T(convR)$. The normed space we seek is then the range of T with the norm given by the Minkowsky functional of $T(convR)$ and the embedding of M into it is given by T.

Let us now describe the details.

Let (M, d) be a metric space with $|M| = m$ points. We shall assume that m is a power of 2, $m = 2^n$. For $0 \leq s \leq m$ put

$$\mathcal{M}_s = \{S \subseteq M;\ |S| = s\}$$

and

$$\mathcal{M} = \cup_{k=0}^{n} \mathcal{M}_{2^k}\ .$$

Let μ_k be the uniform probability measure on $\mathcal{M}_{2^k}$ and let

$$\mu = \frac{1}{n+1} \sum_{k=0}^{n} \mu_k\ .$$

Note that the map $x \to \{d(x, A)\}_{A \in \mathcal{M}}$ is an isometric embedding of (M, d) into $L_\infty(\mathcal{M}, \mu)$ (this embedding was first used in [B]).

The Theorem will follow easily from the next three lemmas. The first is based on Proposition 3 of [B].

LEMMA 1: *For all $0 < \alpha < \frac{1}{3}$, all metric spaces (M, d) as above and all $x, y \in M$,*

$$\mu(\{A \in \mathcal{M};\ |d(x, A) - d(y, A)| \geq \alpha d(x, y)\}) \geq [216(n+1)m^{3\alpha}]^{-1}\ .$$

PROOF: Let $x, y \in M$, $x \neq y$. Put $d = d(x, y)$ and define a sequence of balls around x and y alternatively by

$B_0 = \{x\}$, $B_1 = B(y, \alpha d)$ (= the closed ball with center y and radius αd), $B_2 = B(x, 2\alpha d)$, $B_3 = B(y, 3\alpha d)$ and so on.

Then, for some i with $1 \le i \le \left[\frac{1}{3\alpha}\right] + 1$, $\frac{|B_i|}{|B_{i-1}|} \le m^{3\alpha}$. Note that, since $\alpha < \frac{1}{3}$, $B_i \cap B_{i-1} = \emptyset$. Set $a = |B_i|$, $b = |B_{i-1}|$ and let k be the integer such that $\frac{m}{2} < (a+b)2^k \le m$. We shall show

$$(*) \qquad \mu(\{A \in \mathcal{M};\ A \cap B_i = \emptyset,\ A \cap B_{i-1} \neq \emptyset\}) \ge [216(n+1)m^{3\alpha}]^{-1}$$

This will prove the lemma: Assume for example that i is even then $B_i = B(x, i\alpha d)$, $B_{i-1} = B(y, (i-1)\alpha d)$. If $A \cap B_i = \emptyset$ then $d(x,A) > i\alpha d$. If $A \cap B_{i-1} \neq \emptyset$ then $d(x,A) \le (i-1)\alpha d$. Consequently, if A is in the set in $(*)$ then $|d(x,A) - d(y,A)| \ge \alpha d$.

To prove $(*)$ there is obviously no loss of generality to assume that $b \le a$ (the left hand side of $(*)$ increases if we decrease $|B_{i-1}|$) put $K = 2^k$ and note that

$$\mu_k(\{A \in \mathcal{M};\ A \cap B_i = \emptyset,\ A \cap B_{i-1} \neq \emptyset\}) =$$

$$= \mu_k(\{A \in \mathcal{M}; A \cap B_i = \emptyset\}) - \mu_k(\{A \in \mathcal{M};\ A \cap (B_{i-1} \cup B_i) = \emptyset\})$$

$$= \binom{m}{K}^{-1}\left[\binom{m-a}{K} - \binom{m-a-b}{K}\right]$$

$$= \binom{m}{K}^{-1}\binom{m-a}{K}\left[1 - \frac{(m-a-b)!(m-a-K)!}{(m-a-b-K)!(m-a)!}\right]$$

$$= \binom{m}{K}^{-1}\binom{m-a}{K}\left[1 - \frac{(m-a-b)(m-a-b-1)\cdots(m-a-b-K+1)}{(m-a)(m-a-1)\cdots(m-a-K+1)}\right]$$

$$\ge \binom{m}{K}^{-1}\binom{m-a}{K}\left[1 - \left(\frac{m-a-b}{m-a}\right)^K\right]$$

$$= \binom{m}{K}^{-1}\binom{m-a}{K}\left[1 - \left(1 - \frac{b}{a}\frac{1}{\frac{m}{a}-1}\right)^K\right]$$

$$\ge \binom{m}{K}^{-1}\binom{m-a}{K}\left[1 - \left(1 - \frac{b}{a}\frac{1}{4K-1}\right)^K\right]$$

$$\ge \binom{m}{K}^{-1}\binom{m-a}{K}\left(1 - e^{-\frac{b}{4a}}\right)$$

$$\geq \binom{m}{K}^{-1} \binom{m-a}{K} \frac{b}{8a} \qquad (since\ \frac{b}{4a} \leq \frac{1}{4})$$

$$\geq \left(1 - \frac{K}{m-a}\right)^a \frac{b}{8a} \qquad (this\ is\ a\ similar\ computation)$$

$$\geq \left(1 - \frac{2}{a}\right)^a \frac{b}{8a} \qquad (if\ a > 2,\ otherwise\ \leq \frac{b}{128a})$$

$$\geq \frac{b}{216a} \geq \frac{1}{216 m^{3\alpha}}$$

and $(*)$ follows.

□

We shall consider M as a subset of $L_\infty(\mathcal{M}, \mu)$. Let

$$R = \left\{ \frac{x-y}{d(x,y)};\ x \neq y \quad x, y \in M \right\} \subset L_\infty(\mathcal{M}, \mu)$$

and let $\|\cdot\|_p$ denote the norm in $L_p(\mathcal{M}, \mu)$.

LEMMA 2: *Let $0 < \alpha < \frac{1}{3}$. For each $u \in R$ there exists a $f \in L_2(\mathcal{M}, \mu)$ such that*

(i)
$$< f, u > \ \geq \alpha$$

(ii)
$$| < f, v > | \leq 1 \quad for\ all \quad v \in R$$

(iii)
$$\|f\|_2 \leq 15(n+1)^{1/2} m^{\frac{3\alpha}{2}} .$$

PROOF: Let $u = \frac{x-y}{d(x,y)}$ $x \neq y$. Put

$$\tilde{f}(A) = \begin{cases} 1 & \text{if } |d(x,A) - d(y,A)| \geq \alpha d(x,y) \\ 0 & \text{otherwise} \end{cases}$$

and

$$f = \frac{\tilde{f}}{\|\tilde{f}\|_1} .$$

Then *(i)* and *(ii)* are easily checked and

$$\|f\|_2 = [\mu(\{A;\ |d(x,A) - d(y,A)| \geq \alpha d(x,y)\})]^{-\frac{1}{2}} \leq 15(n+1)^{1/2} m^{\frac{3\alpha}{2}}$$

by Lemma 1.

□

LEMMA 3: *There exists a constant $\delta > 0$ such that if R is a finite subset of the unit ball of ℓ_2^h with $|R| = r$ and $u \in R$ $f \in \ell_2^h$ are such that*

(i)
$$< f, u > \ \geq \alpha > 0 \, ,$$

(ii)
$$| < f, v > | \leq 1 \quad for\ all \quad v \in R \, .$$

Then, if $k \geq \delta^{-1} max\{log\ r, \alpha^{-2}\}(1 + \|f\|^4)$, there exists a rank k orthogonal projection satisfying

(a)
$$< f, Pu > \ > \delta \frac{k}{h} \alpha$$

(b)
$$| < f, Pv > | < 4\delta \frac{k}{h} \qquad for\ all\ v \in R.$$

Moreover, consider the probability measure on the set of rank k projections, obtained from the Haar measure on $U(n)$ via the transformation $U \to U^* P_0 U$ (P_0 some fixed rank k projection). Then (a) and (b) occur with probability $\geq$

$$1 - 8\ exp\left(-\frac{\delta k \alpha^2}{1 + \|f\|^4}\right) - 8r\ exp\left(-\frac{\delta k}{1 + \|f\|^4}\right) \, .$$

PROOF: Let $v \in R$ and for a random projection P of *rank* k let

$$\varphi(P) = < f, Pv > = \frac{1}{4}(\|P(v+f)\|^2 - \|P(v-f)\|^2).$$

The concentration of measure phenomenon on $U(n)$ implies (see the proof of Lemma 1 in [JL] for details) that for some positive constant $c = c(k, h)$, which is bounded away from zero independently of k, h, and for all $w \in \ell_2^h$

$$Prob\left(\left\{P; \left|\|Pw\| - c\sqrt{\frac{k}{h}}\|w\|\right| \geq \varepsilon\sqrt{\frac{k}{h}}\|w\|\right\}\right) \leq 4exp\left(\frac{-k\varepsilon^2}{2}\right) \, .$$

Since, $|A - B| < C$ implies $|A^2 - B^2| \leq (2B + C)C$, we get

$$Prob\left(\left\{P; \left|\|Pw\|^2 - c^2\frac{k}{h}\|w\|^2\right| \geq \varepsilon(2c+\varepsilon)\frac{k}{h}\|w\|^2\right\}\right) \leq 4exp\left(-\frac{k\varepsilon^2}{2}\right) \, .$$

Consequently, with probability $\geq \ 1 - 8\ exp(-\frac{k\varepsilon^2}{2})$, both

$$\left|\|P(v+f)\|^2 - c^2\frac{k}{h}\|v+f\|^2\right| < \varepsilon(2c+\varepsilon)\frac{k}{h}\|v+f\|^2$$

and

$$\left|\|P(v-f)\|^2 - c^2\frac{k}{h}\|v-f\|^2\right| < \varepsilon(2c+\varepsilon)\frac{k}{h}\|v-f\|^2$$

and, with probability $\geq \ 1 - 8exp\left(\frac{-k\varepsilon^2}{2}\right)$

$$(**) \qquad |<f, Pv> - c^2\frac{k}{h} <f, v>| < \varepsilon(2c+\varepsilon)\frac{k}{2h}\left(\|f\|^2 + \|v\|^2\right) .$$

Applying $(**)$ for $v = u$ with $\varepsilon = \varepsilon_0$ satisfying $\frac{\varepsilon}{2}(2c+\varepsilon)(1+\|f\|^2) = \frac{\alpha c^2}{2}$ (so $\varepsilon_0 \geq \frac{\alpha c}{2(1+\|f\|^2)}$), we get

$$Prob\left(\left\{<f, Pu> > \frac{\alpha c^2}{2}\frac{k}{h}\right\}\right) \geq 1 - 8exp\left(\frac{-k\varepsilon_0^2}{2}\right) .$$

Applying $(**)$ for the rest of $v \in R$ with $\varepsilon = \varepsilon_1$ satisfying $\frac{\varepsilon(2c+\varepsilon)}{2}(1+\|f\|^2) = c^2$ (so $\varepsilon_1 \geq \frac{c}{1+\|f\|^2}$), we get

$$Prob\left(\left\{|<f, Pv>| < 2c^2\frac{k}{h} \quad for\ all\ v \in R\right\}\right) \geq 1 - 8r\ exp\left(\frac{-k\varepsilon_1^2}{2}\right) .$$

Combining, we get that with probability larger than or equal to

$$1 - 8exp\left(-\frac{k\alpha^2c^2}{8(1+\|f\|^2)^2}\right) - 8rexp\left(-\frac{kc^2}{2(1+\|f\|^2)^2}\right)$$

we have

$$<f, Pu> > \frac{\alpha c^2}{2}\frac{k}{h} \ \ and \ \ |<f, Pv>| < 2c^2\frac{k}{h} \ \ for\ all\ v \in R .$$

□

PROOF OF THE THEOREM: Put $\alpha = \frac{\beta}{10} < \frac{1}{3}$. Consider M as a subset of $L_\infty(\mathcal{M}, \mu)$, as explained above. Let $R = \left\{\frac{x-y}{d(x,y)};\ x, y \in M,\ x \neq y\right\}$ so that $|R| = r = \frac{m(m-1)}{2}$. By Lemmas 2 and 3 if k is such that

$$(***) \qquad 1 - 8m\ exp\left(-\frac{\delta k\alpha^2}{(n+1)^2m^{6\alpha}}\right) - 8mr\ exp\left(-\frac{\delta k}{(n+1)^2m^{6\alpha}}\right) > 0$$

($\delta > 0$ absolute) then there exists a rank k projection P so that for all $u \in R$ there exists a linear functional $f = f_u$ with

$$(a) \quad < f, Pu > > \delta \frac{k}{m} \alpha \qquad (b) \quad | < f, Pv > | < 4\delta \frac{k}{m} \quad \textit{for all} \quad v \in R \; .$$

It follows that $\frac{4}{\alpha} Pu$ does not belong to $C = conv\{Pv;\ v \in R\}$. Consider the k dimensional normed space X whose unit ball is C, then P maps M into X and

$$\|P_{|M}\|_{Lip} \|(P_{|M})^{-1}\|_{Lip} \leq \frac{4}{\alpha} \; .$$

To evaluate k, notice that $(***)$ is implied by

$$(****) \qquad k \geq K\ max \left\{ \frac{1}{\alpha^2}, n \right\} n^3 m^{6\alpha} \; ,$$

where K is some absolute constant. We shall distinguish between four cases:

(1) For $\alpha > \frac{1}{\sqrt{n}}$,

$$max \left\{ \frac{1}{\alpha^2}, n \right\} n^3 m^{6\alpha} = n^4 m^{6\alpha} \leq m^{7\alpha}$$

for m large enough so that any $k \geq K m^\beta$ is okay.

(2) For $\frac{\log_2 n}{n} \leq \alpha \leq \frac{1}{\sqrt{n}}$,

$$max \left\{ \frac{1}{\alpha^2}, n \right\} n^3 m^{6\alpha} = \frac{1}{\alpha^2} n^3 m^{6\alpha} \leq \frac{n^4}{(\log_2 n)^2} m^{6\alpha} \leq m^{10\alpha}$$

and again any $k \geq K m^\beta$ is okay.

(3) For $\frac{1}{n} \leq \alpha < \frac{\log_2 n}{n}$ we shall leave the estimate $k \geq K \alpha^{-2} n^3 m^\beta$.

(4) For $\alpha < \frac{1}{n}$ one gets a better estimate from the main result of [B]; k can be taken of order $n = \log_2 m$ (and the containing space can be a Hilbert space).

□

We do not know if the Theorem is best possible. The following proposition gives a lower bound in a related problem: What is the smallest dimension of a normed space into which all metric spaces with m points $\frac{1}{\alpha}$-embed?

PROPOSITION 4: *Let $0 < \alpha < 1$ and $m \in \mathbb{N}$. Assume that X is a normed space such that any m-points metric space $\frac{1}{\alpha}$-embeds into X. Then $\dim X \geq c_\alpha m^{\frac{\alpha}{K}}$, where $K > 0$ is a universal constant and $c_\alpha > 0$ depends only on α.*

The proof follows easily from [B] (Lemma 4 and the proof of Proposition 2). Using a similar proof one can also show

PROPOSITION 5: Let $1 < \lambda < \sqrt{2}$, let $m \in \mathbb{N}$ and let X be a normed space such that any m-points metric space λ-embeds into X. Then, for some absolute $\delta > 0$,

$$\dim X \geq \delta \left(\log \frac{1}{\sqrt{2} - \lambda} \right)^{-1} m \, .$$

References

[B] J. Bourgain, On Lipschitz embedding of finite metric spaces in Hilbert space, Israel J. Math. 52 (1985), 46-52.

[JL] W.B. Johnson and J. Lindenstrauss, Extensions of Lipschitz mappings into a Hilbert space, Contemp. Math. 26 (1984), 189 - 206.

RANDOM SERIES IN THE REAL INTERPOLATION SPACES BETWEEN THE SPACES v_p

Gilles PISIER and Quanhua XU

Texas A and M University and Université Paris 6

Wuhan University and Université Paris 6

Introduction

We will study the interpolation spaces between the space of functions of bounded variation and the space of continuous functions. This investigation can be developed for spaces of functions defined on $\mathbb{R}$, $[0,1]$ or $\mathbb{N}$. We will present the results in the framework of sequence spaces (i.e., over $\mathbb{N}$), but we also discuss the case of function spaces at the end of the paper. The motivation of our study has its origin in the work of R.C. James on non-reflexive Banach spaces. For a long time, it was an open question whether a Banach space which does not contain ℓ_1^n's uniformly must be reflexive. Formulated in terms of "type", this question becomes: if X is of type $p > 1$, is X reflexive? James [J1] gave a negative answer. James' construction is simplified in [JL] and in [DL] non-reflexive spaces of type p are constructed for every $p < 2$. Later, James [J2] constructed non-reflexive spaces of type 2.

In this paper, we give a new construction of such examples. Furthermore, we find for every $q > 2$ a non-reflexive space of type 2 and of cotype q. Moreover, we will exhibit a non-reflexive space $\widetilde{X}$ with the following property: there is a $C > 0$ such that for all $n > 1$, all n dimensional subspaces $E \subset \widetilde{X}$ satisfy $d(E, \ell_2^n) \leq C \operatorname{Log} n$.

This is sharp in the following sense: if one replaces $\operatorname{Log} n$ by any function of n which is $o(\operatorname{Log} n)$ then the space must be reflexive (cf. [P1] p. 349, cf. also [Ka]).

These examples are somewhat "natural" function (or sequence) spaces defined by the Lions-Peetre interpolation method and our results give several informations of independent interest on the analysis of functions (or random variables) with values in these spaces.

Let us briefly describe how the spaces are defined.

Let v_1 be the Banach space of all sequences of real number $(\alpha_n)_{n\geq 0}$ of bounded variation, i.e., such that

$$|\alpha_0| + |\alpha_1 - \alpha_0| + \cdots + |\alpha_n - \alpha_{n-1}| + \cdots < \infty \ .$$

Clearly $v_1 \subset \ell_\infty$. Let us denote by i this inclusion map. Among many deep results concerning reflexivity or weak compactness, James proved that a Banach space X is non-reflexive iff the inclusion i factors through X, i.e., we have a factorization

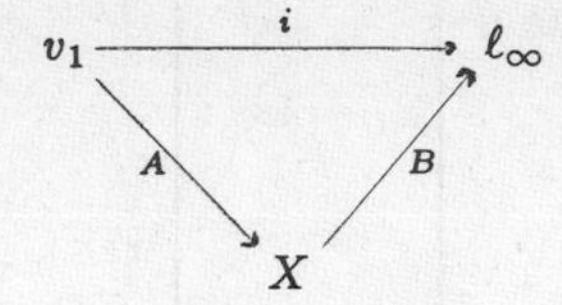

with operators A, B bounded and such that $i = BA$ (cf. [Pt], [J7] and [J6] p. 206 or [LP] p. 322). In other words, every non-reflexive space can be viewed as "intermediate" between v_1 and ℓ_∞. (Since i is not weakly compact, all intermediate spaces are necessarily non-reflexive.) A natural idea is then to consider the interpolation spaces between v_1 and ℓ_∞ and to show that (for a suitable interpolation functor) they possess a good type and cotype. This is what we do in this paper, we study the space $(v_1, \ell_\infty)_{\theta q}$ obtained by the real interpolation method, originally introduced by Lions and Peetre. It is probably interesting to study the complex method of interpolation, we have no significant result in that case yet.

We also give some applications of these methods to the notion of k-structure introduced in [DJL]. We show that for any $q > 2$ there are spaces with a k-structure and admitting type 2 and cotype q. We can also give an (apparently new) application of our results to the theory of bases: every separable Banach space B of type $p > 1$ and cotype $r > 2$ is a quotient of a space of type p and cotype r which has a basis.

Finally, in the last part of the paper, we discuss the connection with "spline approximation". We also discuss the case of function spaces over $[0, 1]$ or $\mathbb{R}$.

Notation and Background

Let B be a Banach space. We will denote by $v_1(B)$ the space of all sequences $x = (x_n)_{n \geq 0}$ of elements of B such that $\|x\|_{v_1(B)} = \|x_0\| + \sum_{n \geq 1} \|x_n - x_{n-1}\| < \infty$.

We denote $v_1(\mathbb{R})$ simply by v_1. This is the space of sequences of bounded variation.

More generally, for $1 \leq p < \infty$, we denote by $v_p(B)$ the space of all sequences $x = (x_n)_{n \geq 0}$ such that the supremum of

$$(1) \qquad (\|x_{n_0}\|^p + \|x_{n_1} - x_{n_0}\|^p + \cdots + \|x_{n_k} - x_{n_{k-1}}\|^p + \cdots)^{1/p}$$

over all increasing sequences $0 \leq n_0 \leq n_1 \leq \cdots$ of integers, is finite. We denote this supremum by $\|x\|_{v_p(B)}$. The space $v_p(B)$ (equipped with this norm) is clearly a Banach space.

It will be useful to replace ℓ_p in (1) by the more general Lorentz spaces ℓ_{pq}. Let us recall the definition of the latter. Let $1 \le p \le \infty$, $1 \le q \le \infty$. Let $\alpha = (\alpha_n)_{n\ge 0}$ be a sequence of scalars and let $(\alpha_n^*)_{n\ge 0}$ be the non-increasing rearrangement of $(|\alpha_n|)_{n\ge 0}$. Then, by definition, α belongs to ℓ_{pq} if $\sum n^{\frac{q}{p}-1}\alpha_n^{*q} < \infty$. We let $\|\alpha\|_{pq} = \left(\sum n^{\frac{q}{p}-1}\alpha_n^{*q}\right)^{1/q}$, and $\|\alpha\|_{p\infty} = \sup n^{1/p}\alpha_n^*$.

Now, if $x = (x_n)_{n\ge o}$ is a sequence in B, we say that x belongs to $v_{pq}(B)$ if the supremum of

$$\left\|(\|x_{n_0}\|, \|x_{n_1} - x_{n_0}\|, \ldots, \|x_{n_k} - x_{n_{k-1}}\|, \ldots)\right\|_{pq},$$

over all sequences (n_k) as above, is finite.

We denote this supremum by $\|x\|_{v_{pq}(B)}$. Clearly, $v_{pp}(B)$ (and its norm) coincides with $v_p(B)$. The space $v_{p\infty}(B)$ is derived from $v_p(B)$ by replacing the ℓ_p norm by the weak-ℓ_p norm.

In the particular case $B = I\!R$, we abbreviate this notation and replace $v_{pq}(I\!R)$ by v_{pq}. We now consider the case $p = \infty$. We will denote as usual $\ell_\infty(B)$ (resp. ℓ_∞) the Banach space of all bounded sequences in B (resp. in $I\!R$), equipped with its natural norm.

We will denote by $c(B)$ (resp. $c_0(B)$) the closed subspace of $\ell_\infty(B)$ formed by all sequences which converge (resp. converge to zero) when n tends to infinity.

If $B = I\!R$, we write simply c and c_0. We have, obviously, an inclusion

$$v_1(B) \subset c(B).$$

Let $v_1^0(B) = v_1(B) \cap c_0(B)$ so that

$$v_1^0(B) \subset c_0(B).$$

The preceding inclusions allow us to consider the interpolation spaces $(v_1(B), \ell_\infty(B))_{\theta q}$ and $(v_1^0(B), c_0(B))_{\theta q}$ obtained by the real interpolation method of Lions-Peetre (cf. [BL] or [BB]). We recall how these spaces can be defined using the so-called K-method.

Let (A_0, A_1) be an interpolation couple of Banach spaces. For $x \in A_0 + A_1$, and $t > 0$, we define $K_t(A_0, A_1; x)$ or more briefly $K_t(x)$ as follows

$$K_t(x) = \inf(\|x_o\|_{A_0} + t\|x_1\|_{A_1} \mid x = x_0 + x_1\}.$$

We denote by $K_t(A_0, A_1)$ the space $A_0 + A_1$ equipped with the norm $K_t(\cdot)$.

Let

$$\|x\|_{\theta q} = \left(\int_0^\infty (t^{-\theta}K_t(x))^q \frac{dt}{t}\right)^{1/q}.$$

The space $(A_0, A_1)_{\theta q}$ is then defined as the space of all x in $A_0 + A_1$ such that $\|x\|_{\theta q} < \infty$ (with the obvious convention for $q = \infty$). Equipped with the norm $\| \ \|_{\theta q}$ it is a Banach space. For example, if $A_0 = \ell_1$ and $A_1 = \ell_\infty$, then $(A_0, A_1)_{\theta q}$ is the Lorentz space ℓ_{pq} with $\frac{1}{p} = \frac{1-\theta}{1} + \frac{\theta}{\infty}$.

We will use interpolation for the couple $\big(L_p(A_0), L_p(A_1)\big)$ relative to a fixed measure space. In that case, it is easy to see that the K_t-norm for $\big(L_p(A_0), L_p(A_1)\big)$ is (uniformly in $t > 0$) equivalent to the norm in $L_p\big(K_t(A_0, A_1)\big)$. Using this observation (and the classical inclusions $L_q(L_p) \supset L_p(L_q)$ if $q \geq p$ and $L_q(L_p) \subset L_p(L_q)$ if $q \leq p$) it is easy to check that

(2) if $q \geq p$ $\quad \big(L_p(A_0), L_p(A_1)\big)_{\theta q} \supset L_p\big((A_0, A_1)_{\theta q}\big)$

(3) if $q \leq p$ $\quad \big(L_p(A_0), L_p(A_1)\big)_{\theta q} \subset L_p\big((A_0, A_1)_{\theta q}\big)$.

This result is well known to specialists in interpolation theory.

Besides the spaces $v_p(B)$, it will be convenient to introduce the spaces $u_p(B)$ which are essentially preduals of the spaces $v_p(B)$. Here is how we proceed. Fix $1 < p < \infty$. Let S be the space of all sequences $x = (x_n)_{n \geq 0}$ in B which are constant for n large enough. Let $x \in S$. We consider all representations of x of the following form:

$$x_n = \sum_{k=1}^{N} y_k 1_{I_k}(n) \tag{4}$$

where $I_1, I_2, \ldots, I_N$ are disjoint consecutive intervals of integers which cover $\mathbb{N}$ where y_k belongs to B, and $y_k \neq y_{k+1}$ for all $k \geq 1$. Let then $[\![x]\!] = \left(\sum \|y_k\|^p\right)^{1/p}$. Now, we consider all decompositions of x in the form $x = \sum_1^N z^k$ and we let

$$\|x\|_{u_p(B)} = \inf \left\{ \sum_1^N [\![z^k]\!] \right\}$$

where the infimum runs over all such representations. The space $u_p(B)$ is then defined as the completion of S for the norm $u_p(B)$. It is rather easy to check that

$$\|x\|_{\ell_\infty(B)} \leq \|x\|_{u_p(B)} \leq \|x\|_{v_1(B)}$$

and we have $v_1(B) \subset u_p(B) \subset \ell_\infty(B)$. In fact $u_p(B) \subset c(B)$. The definition of $u_p(B)$ is adjusted so that one can check rather easily that the dual of $u_p(B)$ is naturally identifiable with $v_{p'}(B^*)$, $\frac{1}{p} + \frac{1}{p'} = 1$. We may note also that $u_p(B) \subset v_p(B)$.

We will need to recall some notation and definitions relative to type and cotype.

Let $D = \{-1,+1\}^{\mathbb{N}}$ equipped with its standard probability measure μ. We denote by $\varepsilon_n : D \to \{-1,+1\}$ the n-th coordinate on D. Let us recall the classical Khintchine inequalities. For each $0 < p < \infty$, there are constants $A_p > 0$ and B_p such that

$$\text{(5)} \qquad \forall (\alpha_n) \in \ell_2 \quad A_p \left(\sum |\alpha_n|^2\right)^{1/2} \leq \|\sum \alpha_n \varepsilon_n\|_{L_p(\mu)} \leq B_p \left(\sum |\alpha_n|^2\right)^{1/2}$$

A Banach space B is called of type p (resp. of cotype q) if there is a constant C such that for all finite sequences (x_i) in B we have

$$\|\sum \varepsilon_i x_i\|_{L_2(\mu;B)} \leq C \left(\sum \|x_i\|^p\right)^{1/p}$$

(resp. $\geq \frac{1}{C}(\sum \|x_i\|^q)^{1/q}$).

We denote by $T_p(B)$ (resp. $C_q(B)$) the smallest constant C for which this holds. We refer to [MP] for more details on these notions.

We will prove the following

Theorem 1: Let $0 < \theta < 1$, $1 \leq q < \infty$. Define $p_\theta = (1-\theta)^{-1}$ so that $\frac{1}{p_\theta} = \frac{1-\theta}{1} + \frac{\theta}{\infty}$.

(i) If $\theta \neq 1/2$, the space $(v_1, \ell_\infty)_{\theta q}$ is of type $p_\theta \wedge q \wedge 2$ and of cotype $p_\theta \vee q \vee 2$.

(ii) If $\theta = 1/2$, then $(v_1, \ell_\infty)_{\frac{1}{2} q}$ is of type $q \wedge (2-\varepsilon)$ and of cotype $q \vee (2+\varepsilon)$ for every $\varepsilon > o$.

(iii) Moreover, there is a constant C such that, for all finite dimensional subspaces $E \subset (v_1, \ell_\infty)_{\frac{1}{2} 2}$ with $\dim E = n > 1$, we have

$$d(E, \ell_2^n) \leq C \operatorname{Log} n .$$

Remark: It is rather easy to see that $(v_1, \ell_\infty)_{\frac{1}{2} 2}$ is isomorphic to its dual, and more generally that $(v_1, \ell_\infty)^*_{\theta q}$ is isomorphic to $(v_1, \ell_\infty)_{1-\theta\ q'}$ for $q < \infty$ and $0 < \theta < 1$. Indeed, note that since $v_1 \subset c$ we have $(v_1, \ell_\infty)_{\theta q} = (v_1, c)_{\theta q}$ and hence

$$(v_1, \ell_\infty)^*_{\theta q} = (v_1^*, c^*)_{\theta q'} .$$

But the Abel transformation $(\xi_n)_n \to (\xi_0, \xi_0 + \xi_1, \xi_0 + \xi_1 + \xi_2, \ldots)$ essentially identifies v_1^* with $\ell_\infty \oplus \mathbb{R}$ and c^* with $v_1 \oplus \mathbb{R}$. Hence, we obtain

$$\begin{aligned}
(v_1, \ell_\infty)^*_{\theta q} &\approx (v_1 \oplus \mathbb{R}, \ell_\infty \oplus \mathbb{R})_{1-\theta\ q'} \\
&\approx (v_1, \ell_\infty)_{1-\theta\ q'} \oplus \mathbb{R} \\
&\approx (v_1, \ell_\infty)_{1-\theta\ q'} ,
\end{aligned}$$

which proves the above claim.

We also should mention that it is easy to prove (cf. [DFJP]) that $(v_1, \ell_\infty)_{\theta q}$ is of codimension one in its bidual when $0 < \theta < \infty$ and $1 \le q < \infty$. Thus, we obtain a new proof of some results in the paper [J5].

The next result from [BP] (cf. also [Kr], [KS]) will be crucial in the sequel. We include a proof for the sake of completeness.

Lemma 2: Let $0 < \theta < 1$, $1 \le q \le \infty$, and let $p_\theta = (1-\theta)^{-1}$ so that $\frac{1}{p_\theta} = \frac{1-\theta}{1} + \frac{\theta}{\infty}$. Then, for any Banach space B, we have

(i) $\big(v_1(B), \ell_\infty(B)\big)_{\theta q} \subset v_{p_\theta q}(B)$,

(ii) $v_{p_\theta \infty}(B) \subset \big(v_1(B), \ell_\infty(B)\big)_{\theta\infty}$,

(iii) $\big(v_1(B), \ell_\infty(B)\big)_{\theta 1} \subset u_{p_\theta}(B) \subset \big(v_1(B), \ell_\infty(B)\big)_{\theta p_\theta}$.

Moreover, the norms of all these inclusions are less than a constant independent of B.

Proof: Part (i) is easy. We briefly sketch the argument. Let $n_0 \le n_1 \le \dots$ be any increasing sequence of integers. Consider the operator $u : v_1(B) \to \ell_1(B)$ defined by

$$u(x) = (x_{n_0}, x_{n_1} - x_{n_0}, \dots, x_{n_k} - x_{n_{k-1}}, \dots) \,.$$

This operator has norm not more than 1, but it also has norm not more than 2 from $\ell_\infty(B)$ into itself, therefore it is bounded from $\big(v_1(B), \ell_\infty(B)\big)_{\theta q}$ into $\ell_{p_\theta q}(B) = \big(\ell_1(B), \ell_\infty(B)\big)_{\theta q}$.

Let us prove part (ii). Let x be such that $\|x\|_{v_{p\infty}(B)} \le 1$ and let $t > 0$.

Let us denote $K_t(x)$ instead of $K_t\big(v_1(B), \ell_\infty(B); x\big)$. We will show that $K_t(x) \le C'' t^\theta$ for $\theta = 1 - \frac{1}{p}$ and some constant C''. Note that the case $t \le 1$ is trivial here since $K_t(x) \le t\|x\|_{\ell_\infty(B)} \le t \le t^\theta$.

Let us assume $t > 1$.

Then let $n_1 = \inf\{n > 0 \mid \|x_n - x_0\| \ge t^{-(1-\theta)}\}$. We then define similarly $n_2, n_3, \dots$ and $n_k = \inf\{n > n_{k-1} \mid \|x_n - x_{n_{k-1}}\| \ge t^{-(1-\theta)}\}$.

Of course, the preceding "infimum" may run over the void set, in which case we let $n_k = \infty$ and we stop the process. Clearly since $\|x\|_{v_{p\infty}(B)} \le 1$, we have

$$\begin{aligned}\|x_{n_1} - x_0\| + \dots + \|x_{n_k} - x_{n_{k-1}}\| &\le 1 + \dots + k^{-1/p} \\ &\le C k^{1/p'}\end{aligned}$$

for some constant $C = C(p)$. Hence if $n_1, \dots, n_k$ are all finite, we must have

$$k t^{-(1-\theta)} \le C k^{1/p'}$$

or equivalently $k \leq C^p t$.

This shows that k cannot be too large.

We now assume that n_k is the last finite integer in the sequence $n_1 < n_2 < \ldots$.

Let

$$I_0 = [0, n_1[\ , I_1 = [n_1, n_2[\ , \ldots\ldots ,$$

$$I_k = [n_k, \infty[\ .$$

We set $n_0 = 0$. Then for any interval I_j in the preceding list we have clearly

$$\sup_{n \in I_j} \|x_n - x_{n_j}\| < t^{-(1-\theta)} .$$

Hence we may decompose x as $x = x^0 + x^1$ with

$$x_n^0 = x_{n_j} \quad \text{if} \quad n \in I_j$$

and

$$x_n^1 = x_n - x_n^0 .$$

We have then

$$\|x^1\|_{\ell_\infty(B)} \leq t^{-(1-\theta)}$$

and

$$\begin{aligned} \|x^0\|_{v_1(B)} &\leq \|x_0\| + \sum_{j=1}^{k} \|x_{n_j} - x_{n_{j-1}}\| \\ &\leq \|x_0\| + Ck^{1/p'} \\ &\leq t^\theta (C^{p/p'+1} + 1) . \end{aligned}$$

Hence

$$\begin{aligned} K_t(x) &\leq \|x^0\|_{v_1(B)} + t\|x^1\|_{\ell_\infty(B)} \\ &\leq C'' t^\theta \quad \text{for some constant } C'' \text{ .q.e.d.} \end{aligned}$$

The last part (iii) is essentially a dualization of the inclusions

$$\big(v_1(B), \ell_\infty(B)\big)_{\theta p_\theta} \subset v_{p_\theta}(B) \subset \big(v_1(B), \ell_\infty(B)\big)_{\theta \infty} .$$

It can be justified as follows. First consider x in S such that (4) holds. Since the norm of (y_k) in $\ell_{p_\theta}(B)$ is equivalent to its norm in $\big(\ell_1(B), \ell_\infty(B)\big)_{\theta p_\theta}$, it is easy to obtain

$$\|x\|_{(v_1(B), \ell_\infty(B))_{\theta p_\theta}} \leq C[\![x]\!]$$

for some constant C.

By an obvious convexity argument, we may replace $[\![x]\!]$ by $\|x\|_{u_{p_\theta}(B)}$ in the last inequality and we obtain the right side of (iii).

The other side follows easily from (ii) using the duality $u_p(B)^* \approx v_{p'}(B^*)$ and $v_{p'}(B^*) \subset \left(v_1(B^*), \ell_\infty(B^*)\right)_{1-\theta\ \infty}$. We leave the details to the reader.

Remark: Clearly the same proof shows that $\left(v_1^0(B), c_0(B)\right)_{\theta q} \subset v_{pq}^0(B)$ and (by an easy modification)

$$v_{p\infty}^0(B) \subset \left(v_1^0(B), c_0(B)\right)_{\theta\infty} .$$

Before we go on, it is worthwhile to recall several facts about the type and cotype of Banach lattices, since we will use an analogous approach. Let L be a Banach lattice which is p-convex and r-concave in the sense of [LT2]. This means that $\ell_p(L) \subset L(\ell_p)$ and $L(\ell_r) \subset \ell_r(L)$. It follows immediately that if T is a bounded operator from ℓ_p into ℓ_r (or more generally from L_p into L_r) then T defines canonically a *bounded* operator from $\ell_p(L)$ into $\ell_r(L)$ (resp. from $L_p(L)$ into $L_r(L)$). From such properties, the type and cotype of L can easily be derived. Although the spaces $(v_1, \ell_\infty)_{\theta q}$ are *not* Banach lattices (and actually they even fail l.u.st. cf. [JT]), we will see below that they have properties very similar to the p-convexity or r-concavity of lattices. More precisely, we have

Theorem 3: Let B be a Banach space, $0 < \theta < 1$, $1 \le q \le \infty$. We will denote by $A_{\theta q}(B)$ the space

$$\left(v_1(B), \ell_\infty(B)\right)_{\theta q} .$$

Let p_θ be determined by $\frac{1}{p_\theta} = \frac{1-\theta}{1} + \frac{\theta}{\infty}$. Let (Ω, m) be any measure space. Let us denote simply $L_r(B)$ for $L_r(\Omega, m; B)$. Then

(i) If $p < p_\theta$ and $p \le q$, the following bounded inclusion holds

$$L_p\left(A_{\theta q}(B)\right) \subset A_{\theta q}\left(L_p(B)\right) .$$

(ii) If $r > p_\theta$ and $r \ge q$, then

$$A_{\theta q}\left(L_r(B)\right) \subset L_r\left(A_{\theta q}(B)\right) .$$

Moreover, the norms of these inclusions are majorized by constants depending only on p, q, r and θ.

Proof: (i) The proof is based on the reiteration theorem. By Lemma 2, if $1 \le p < p_\theta$ and $1/p = 1 - \theta'$, we have (since $v_p = v_{pp} \subset v_{p\infty}$)

$$\left(v_1(B), \ell_\infty(B)\right)_{\theta' p} \subset v_p(B) \subset \left(v_1(B), \ell_\infty(B)\right)_{\theta' \infty} .$$

By the reiteration theorem (cf. [BL] p. 50 or [BB] p. 178) this implies that

$$A_{\theta q}(B) = \left(v_p(B), \ell_\infty(B)\right)_{\eta q} \tag{6}$$

with η defined by the relation $(1-\eta)\theta' + \eta = \theta$ (or equivalently $\frac{1-\eta}{p} = \frac{1-\theta}{1}$).

Now, since (6) is valid for an arbitrary B, we may replace B by $L_p(B)$ and we find

$$A_{\theta q}(L_p(B)) = \Big(v_p(L_p(B)), \ell_\infty(L_p(B))\Big)_{\eta q} .$$

But the inclusions $L_p(\ell_\infty(B)) \subset \ell_\infty(L_p(B))$ and $L_p(v_p(B)) \subset v_p(L_p(B))$ are easily checked to be of norm ≤ 1, so that by interpolation, we have an inclusion

$$\Big(L_p(\ell_\infty(B)), L_p(v_p(B))\Big)_{\eta q} \subset A_{\theta q}(L_p(B)) .$$

Now, if $q \ge p$, we have by (2)

$$L_p\Big(\left(v_p(B), \ell_\infty(B)\right)_{\eta q}\Big) \subset \Big(L_p(v_p(B)), L_p(\ell_\infty(B))\Big)_{\eta q} .$$

Hence using (6) once again, we obtain the first part of Theorem 3.

The second part could be proved by invoking a suitable duality argument. One can also repeat an argument similar to the preceding one, as follows. We again invoke the reiteration theorem (and Lemma 2 (iii)) to justify

$$A_{\theta q}(B) = \left(v_1(B), u_r(B)\right)_{\eta q} \tag{7}$$

where $r > p_\theta$ and $1 - \eta + \frac{\eta}{r} = \frac{1}{p_\theta}$.

As above, we use (7) not only for B but also for $L_r(B)$ so that

$$A_{\theta q}(L_r(B)) = \Big(v_1(L_r(B)), u_r(L_r(B))\Big)_{\eta q} .$$

Now going back to the definition of $[\![\quad]\!]$ and $u_r(L_r(B))$ it is easy to check that

$$u_r(L_r(B)) \subset L_r(u_r(B)) .$$

On the other hand, we have obviously

$$v_1(L_r(B)) \subset L_r(v_1(B)) \ .$$

Therefore, we obtain

$$A_{\theta q}(L_r(B)) \subset \Big(L_r(v_1(B)), L_r(u_r(B))\Big)_{\eta q}$$

hence by (3) if $r \geq q$

$$\subset L_r\Big(\big(v_1(B), u_r(B)\big)_{\eta q}\Big) \ .$$

Finally, using (7) again we obtain the second part of Theorem 3. q.e.d.

Remark 4: Let us record here an easy observation concerning the spaces $A_{\theta q}(B)$. Let $T : B_1 \to B_2$ be a bounded operator between Banach spaces.

Then T (or more rigorously $I_{A_{\theta q}} \otimes T$) naturally extends to a bounded operator from $A_{\theta q}(B_1)$ into $A_{\theta q}(B_2)$.

Indeed this is clear for $v_1(B_i)$ and $\ell_\infty(B_i)$ $(i = 1, 2)$ hence it follows for $A_{\theta q}(B_i)$ also by interpolation.

Corollary 5: Let p, q, r and p_θ be as in Theorem 3. Let X be a Banach space and let $T : L_p \to X$ (resp. $T : X \to L_r$) be a bounded operator. Then the operator $T \otimes I_{A_{\theta q}}$ is bounded from $L_p(A_{\theta q})$ into $A_{\theta q}(X)$ (resp. from $A_{\theta q}(X)$ into $L_r(A_{\theta q})$) and the corresponding norms satisfy

$$\|T \otimes I_{A_{\theta q}}\| \leq C\|T\|$$

for some constant C depending only on q, θ and p (resp. and r).

Proof: By Theorem 3, we have $L_p(A_{\theta q}) \subset A_{\theta q}(L_p)$ and by Remark 4, T defines a bounded operator from $A_{\theta q}(L_p)$ into $A_{\theta q}(X)$, hence the desired result follows by composition.

The proof of the part in parenthesis is entirely similar.

Remark: Let X be a Banach space and let $T : L_p \to L_r$ be a bounded operator (L_p and L_r are relative to arbitrary measure spaces). Then the operator $T \otimes I_X : L_p \otimes X \to L_r \otimes X$ is well defined by $T \otimes I_X(\varphi \otimes x) = T(\varphi) \otimes x$ but it does not necessarily define a bounded operator from $L_p(X)$ into $L_r(X)$. In some cases however this is true as the next result shows. In that case we often abusively write T instead of $T \otimes I_X$.

Corollary 6: With the notation of Theorem 3 ($p < p_\theta < r$ and $p \leq q \leq r$), let (Ω_1, m_1) and (Ω_2, m_2) be measure spaces and consider a bounded operator $T : L_p(m_1) \to L_r(m_2)$ such that

$T\otimes I_B$ is bounded from $L_p(B)$ into $L_r(B)$. Then $T\otimes I_{A_{\theta q}(B)}$ is bounded from $L_p\big(A_{\theta q}(B)\big)$ into $L_r\big(A_{\theta q}(B)\big)$. Moreover, we have

$$\|T\otimes I_{A_{\theta q}(B)}\|\le C\|T\otimes I_B\|$$

for some constant C depending only on p,r and θ.

Proof: This is an immediate consequence of Theorem 3. Indeed, it is clear by Remark 4 that T is bounded from $A_{\theta q}\big(L_p(B)\big)$ into $A_{\theta q}\big(L_r(B)\big)$ hence, by Theorem 3, T is bounded from $L_p\big(A_{\theta q}(B)\big)$ into $L_r\big(A_{\theta q}(B)\big)$. q.e.d

Remark: Let $1<p<p'<\infty$, $\frac{1}{p}+\frac{1}{p'}=1$. Let G by any locally compact Abelian group. Then the Fourier transform $\mathcal{F}$ is bounded from $L_p(G)$ into $L_{p'}(\widehat{G})$. By Corollary 6 (applied with $B=\mathbb{R}$ or C), $\mathcal{F}$ is also bounded from $L_p(G;A_{\theta 2})$ into $L_{p'}(\hat{G};A_{\theta 2})$ if $p<p_\theta<p'$. To our knowledge, this is the first *non-reflexive* example of such a phenomenon.

Corollary 7: Let B be of type $p>1$ and of cotype r. Then, if $p<p_\theta<r$ and $1<q<\infty$, the space $A_{\theta q}(B)$ is of type $p\wedge q$ and of cotype $r\vee q$.

Proof: Let $D=\{-1,+1\}^{\mathbb{N}}$ equipped with the standard probability measure μ. Choose $s>q\vee p_\theta$. Let $\overline{p}=p\wedge q$. If B is of type p, the operator $T:\ell_{\overline{p}}\to L_s$ defined by $(\alpha_n)\to\sum\varepsilon_n\alpha_n$ is bounded from $\ell_{\overline{p}}(B)$ into $L_s(B)$. By Corollary 6, the operator $T\otimes I_{A_{\theta q}(B)}$ is bounded from $\ell_{\overline{p}}\big(A_{\theta q}(B)\big)$ into $L_s\big(A_{\theta q}(B)\big)$, which equivalently means that $A_{\theta q}(B)$ is of type $\overline{p}$. For the cotype, we can proceed similarly considering the operator $T:L_{\overline{p}}(\mu)\to\ell_{r\vee q}$ defined by $T(f)=(\int f\varepsilon_n\,d\mu)_{n\in\mathbb{N}}$ for $f\in L_{\overline{p}}$.

By known results, since B is K-convex (cf. [P2]), T is bounded from $L_p(B)$ into $\ell_{r\vee q}(B)$, by Corollary 6 the same holds for $A_{\theta q}(B)$ so that $A_{\theta q}(B)$ is of cotype $r\vee q$. q.e.d.

To put the next result in the right perspective, the reader should recall that if L is a Banach lattice which is q-concave for some $q<\infty$, then for $x_1,\ldots,x_n$ in L we have

$$\|\sum\varepsilon_i x_i\|_{L_2(L)}\approx\|(x_1,\ldots,x_n)\|_{L(\ell_2^n)}$$

and the constants involved in this equivalence are independent of n. This is originally due to Maurey (cf. [LT2] p. 50).

It turns out that an analogous result holds for the spaces $A_{\theta q}$ as follows.

Theorem 8: Let $0<\theta<1$, $1\le q<\infty$. There is a constant $C=C(\theta,q)$ such that, for all finite sequences $x_1,\ldots,x_n$ in $A_{\theta q}$, we have

$$\begin{aligned}\frac{1}{C}\|(x_1,\ldots,x_n)\|_{A_{\theta q}(\ell_2^n)}&\le\|\sum\varepsilon_i x_i\|_{L_2(A_{\theta q})}\\&\le C\|(x_1,\ldots,x_n)\|_{A_{\theta q}(\ell_2^n)}\,.\end{aligned}\tag{8}$$

Proof: We first prove the left side of (8) for $q > 1$. This is an easy consequence of Corollary 5 and the classical Khintchine inequalities (5). Let $1 < p < \infty$ and let $T : L_p(\mu) \to \ell_2^n$ be the operator defined by

$$\forall f \in L_p(\mu) \qquad T(f) = \left(\int f\varepsilon_i \, d\mu\right)_{i=1,2,\dots,n} . \tag{9}$$

Then, by Khintchine's inequalities we have $\|T\| \le C_p$ for some constant C_p. Let $\widetilde{T} = T \otimes I_{A_{\theta q}}$. By Corollary 5 (choosing of course $p < p_\theta \wedge q$) we obtain for any f in $L_p(A_{\theta q})$

$$\|\widetilde{T} f\|_{A_{\theta q}(\ell_2^n)} \le C' \|f\|_{L_p(A_{\theta q})} \tag{10}$$

for some constant C' (independent of n and f). In particular, taking $f = \sum \varepsilon_i x_i$ (and assuming $p \le 2$) we obtain the left side of (8). To prove the right side, we use the second part (in parenthesis) of Corollary 5 applied to the operator $T : \ell_2^n \to L_r(\mu)$ defined by

$$\forall \alpha = (\alpha_i) \in \ell_2^n \qquad T\alpha = \sum_1^n \alpha_i \varepsilon_i .$$

Using Khintchine's inequalities again and choosing $r \ge q$, $r > p_\theta$ and $r \ge 2$, the right side of (8) follows immediately.

To check the left side of (8) for $q = 1$, we recall a classical inequality due to Kahane (cf. e.g., [LT2] p. 74). There is a constant K such that , for all Banach spaces B and for all finite sequences (y_i) in B, we have:

$$\|\sum \varepsilon_i y_i\|_{L_2(B)} \le K \|\sum \varepsilon_i y_i\|_{L_1(B)} . \tag{11}$$

We first invoke the reiteration theorem to justify that we have

$$A_{\theta 1} = (A_0, A_1)_{\theta 1}$$

for $A_0 = A_{\theta_0 2}$ and $A_1 = A_{\theta_1 2}$ with θ_0, θ_1 suitably chosen. We will then apply (11) to the normed space $K_t(A_0, A_1)$ (i.e., $A_0 + A_1$ equipped with the K_t-norm) which we denote simply by K_t. This yields for all x_i in K_t

$$\|\sum \varepsilon_i x_i\|_{L_2(K_t)} \le K \|\sum \varepsilon_i x_i\|_{L_1(K_t)} . \tag{12}$$

Using the observation that the norm of $L_p(K_t)$ is (uniformly in t) equivalent to the norm in $K_t\big(L_p(A_0), L_p(A_1)\big)$ both for $p = 1$ and $p = 2$, we then deduce from (12) after a suitable integration

$$\|\sum \varepsilon_i x_i\|_{(L_2(A_0),L_2(A_1))_{\theta 1}} \le K'\|\sum \varepsilon_i x_i\|_{L_1(A_{\theta 1})}\,, \tag{13}$$

for some constant K'.

Now by (10) $\widetilde{T}$ is uniformly bounded both from $L_2(A_i)$ into $A_i(\ell_2^n)$ for $i = 0, 1$. Hence, by interpolation we have

$$\|\widetilde{T}f\|_{A_{\theta 1}(\ell_2^n)} \le K''\|f\|_{(L_1(A_0),L_2(A_1))_{\theta 1}} \tag{14}$$

for all f in $\big(L_2(A_0), L_2(A_1)\big)_{\theta 1}$. Applying (14) with $f = \sum \varepsilon_i x_i$ and using (13), we finally obtain the left side of (8) for $q = 1$.

Remark 9: For a Banach space X, the space Rad (X) is defined as the closure in $L_2(D, \mu; X)$ of the subspace formed by all elements of the form $\sum \varepsilon_i x_i$ with $x_i \in X$. The above inequality (8) says that Rad $(A_{\theta q})$ can be identified with $A_{\theta q}(\ell_2)$. More generally, if B is of type p for some $p > 1$, the same argument shows (using the notion of K-convexity, cf [P2]) that Rad $\big(A_{\theta q}(B)\big)$ can be identified with $A_{\theta q}\big(\text{Rad }(B)\big)$ if $0 < \theta < 1$ and $1 < q < \infty$.

We will also need the following simple variant of Theorem 3.

Lemma 10: Let B be a Banach space. Assume $1 \le q \le \infty$.

(i) If $p_\theta < t \le \infty$ and $s = p_\theta \wedge q$, then

$$\ell_s\big(A_{\theta q}(B)\big) \subset A_{\theta q}\big(\ell_t(B)\big)\,. \tag{15}$$

(ii) If $1 \le t < p_\theta$ and $r = p_\theta \vee q$, then

$$A_{\theta q}\big(\ell_t(B)\big) \subset \ell_r\big(A_{\theta q}(B)\big)\,. \tag{16}$$

Proof: For simplicity, we assume $B = \mathbb{R}$ but the same argument works in the general case. Also, we will prove only (15). (The second part follows from either a duality argument or a similar argument to the following one.)

By the preceding discussion, we have

$$A_{\theta q} = (v_1, v_t)_{\eta q} \quad \text{with} \quad \frac{1}{p_\theta} = \frac{1-\eta}{1} + \frac{\eta}{t}\,.$$

Moreover $\ell_1(v_1) \approx v_1(\ell_1) \subset v_1(\ell_t)$ and $\ell_t(v_t) \subset v_t(\ell_t)$, hence

$$\big(\ell_1(v_1), \ell_t(v_t)\big)_{\eta q} \subset \big(v_1(\ell_t), v_t(\ell_t)\big)_{\eta q} = A_{\theta q}(\ell_t) \ .$$

Now, if $s = p_\theta \wedge q$, it is easy to check (for more details see e.g., [Mi]) that

$$\ell_s\big((v_1, v_t)_{\eta q}\big) \subset \big(\ell_1(v_1), \ell_t(v_t)\big)_{\eta q} \ ,$$

so that we obtain (15). q.e.d.

In the limiting case $t = p_\theta$, none of the inclusions in Lemma 10 is bounded in general, but the following statement holds. For simplicity, we restrict ourselves to the case $p_\theta = q = 2$.

Lemma 11:

There is a numerical constant C such that, for any Banach space B and any $n > 1$, for every $x = (x_1, \ldots, x_n)$ in $\ell_2^n\big(A_{\frac{1}{2}2}(B)\big)$, we have

$$(17) \qquad C^{-1}(\mathrm{Log}\, n)^{-1/2}\|x\|_{\ell_2^n(A_{\frac{1}{2}2}(B))} \le \|x\|_{A_{\frac{1}{2}2}(\ell_2^n(B))} \le C(\mathrm{Log}\, n)^{1/2}\|x\|_{\ell_2^n(A_{\frac{1}{2}2}(B))} \ .$$

Proof: We will use the same argument as in Lemma 10 but this time we let $t \to 2$ (hence $\eta \to 0$) and we take into account the behaviour of the constants involved in (15) and (16) when $t \to p_\theta = 2$. Recall that if $\theta = 1 - \frac{1}{t}$ we have by Lemma 2

$$(18) \qquad A_{\theta t}\big(\ell_2^n(B)\big) \subset v_t\big(\ell_2^n(B)\big) \subset A_{\theta\infty}\big(\ell_2^n(B)\big)$$

and the norms of these inclusions remain uniformly bounded in n when $t \to 2$.

To simplify, we will show the proof of the right side of (17) only. The proof of the other side is similar. We also assume $B = \mathbb{R}$ to simplify the notation. Then we can use the following chain of inequalities for $x = (x_1, \ldots, x_n)$ in $A_{\frac{1}{2}2}(\ell_2^n)$, and $t < 2$, $\theta = 1 - \frac{1}{t}$, $1 - \eta = t/2$ (so that $\frac{1}{t} = \frac{1-\theta}{1} + \frac{\theta}{\infty}$ and $\frac{1}{2} = \frac{1-\eta}{t} + \frac{\eta}{\infty}$). Note that $\eta \to 0$ when $t \to 2$.

$$\begin{aligned}
\|x\|_{A_{\frac{1}{2}2}(\ell_2^n)} &\le C_1\|x\|_{(A_{\theta\infty}(\ell_2^n), \ell_\infty(\ell_2^n))_{\eta 2}} \\
&\le C_1C_2\|x\|_{(v_t(\ell_2^n), \ell_\infty(\ell_2^n))_{\eta 2}} \\
&\le C_1C_2\|x\|_{(\ell_t^n(v_t), \ell_t^n(\ell_\infty))_{\eta 2}} \\
&\le C_1C_2C_3\|x\|_{\ell_t^n((v_t, \ell_\infty)_{\eta 2})}
\end{aligned}$$

hence using (18)

$$\begin{aligned}
&\le C_1C_2C_3C_4\|x\|_{\ell_t^n((v_1, \ell_\infty)_{\theta 2})} \\
&\le C_1C_2C_3C_4 n^{\frac{1}{t}-\frac{1}{2}}\|x\|_{\ell_2^n(A_{\frac{1}{2}2})} \ .
\end{aligned}$$

The constants C_1, C_2, C_3, C_4 depend on the parameter t. Let us examine their behaviour when $t \to 2$. The constants C_1 and C_4 are very precisely majorized in Holmsted's paper [H]. His estimates ([H], Theorem 3.1 and Remark 3.2) yield

$$C_1 \leq K \quad \text{and} \quad C_4 \leq K\eta^{-1/2}$$

for some constant K.

On the other hand a careful checking of C_2 and C_3 (cf. Lemma 2.(ii) and inequality (2)) shows that C_2 and C_3 remain bounded when $t \to 2$. This yields finally

$$\|x\|_{A_{\frac{1}{2}2}(\ell_2^n)} \leq K\eta^{-1/2} n^{\eta/t} \|x\|_{\ell_2^n(A_{\frac{1}{2}2})}$$

hence choosing $\eta = (\text{Log } n)^{-1}$ we obtain the right side of (17). q.e.d.

Proof of Theorem 1:

We start with part (i). We may as well assume that $p_\theta < 2$ (the case $p_\theta > 2$ is similar). Then Theorem 8 combined with (15) implies that $A_{\theta q}$ is of type $p_\theta \wedge q$. On the other hand, it follows from Theorem 8 and Theorem 3(ii) that (if $p_\theta < 2$) $A_{\theta q}$ is of cotype $2 \vee q$. This completes the proof of part (i).

If $q > 1$, the second part (ii) follows immediately from Corollary 7 applied with B one dimensional. For $q = 1$, it follows from Theorem 8 and Theorem 3(ii). To prove part (iii), let t_n be the type 2 constant of the space $A_{\frac{1}{2}2}$ restricted to n-tuples of vectors. It follows from Theorem 8 and Lemma 11 that

$$t_n \leq K(\text{Log } n)^{1/2}$$

for some constant K.

By known results (cf. [K] and [TJ]) (since $A_{\frac{1}{2}2}$ is isomorphic to its dual) there is a numerical constant K_1 such that for all $n > 1$ and all n-dimensional subspaces $E \subset A_{\frac{1}{2}2}$ we have

$$d(E, \ell_2^n) \leq K_1 t_n^2 \leq K_1 K^2 (\text{Log } n) .$$

This yields the third part of Theorem 1. q.e.d

Remark: The proofs of (8) for $q = 1$ and for Theorem 1 for $q = 1$ are inspired by those of the following result from [X1]. If (A_0, A_1) is an interpolation couple and if A_1 is K-convex and of cotype q_1 then the space $(A_0, A_1)_{\theta 1}$ is of cotype q for $\frac{1}{q} = \frac{1-\theta}{\infty} + \frac{\theta}{q_1}$. The latter result is a generalization of the fact that the Lorentz space L_{q1} is of cotype $q \vee 2$ if $q \neq 2$ and of cotype $2+\varepsilon$ for all $\varepsilon > 0$ if $q = 2$.

Remark: With the notation of this paper, the famous James space J is the space v_2. In connection with Theorem 1, we should recall that it is apparently an open problem whether the space J^* is of finite cotype (or even of cotype 2). Indeed, although J itself (alias v_2) does not seem to be identifiable with a "natural" interpolation space between v_1 and ℓ_∞, it turns out that $(v_1, \ell_\infty)_{\frac{1}{2}\infty}$ coincides with $v_{2\infty}$, and by Theorem 1 the predual (and hence also the dual) of $v_{2\infty}$ is of cotype $2+\varepsilon$ for all $\varepsilon > 0$. In other words, if we replace in the definition of J the ℓ_2-norm by the weak-ℓ_2 norm, then the dual of the resulting space is of cotype $2+\varepsilon$ for all $\varepsilon > 0$. (The fact that $v_{p\infty} = (v_1, \ell_\infty)_{\theta\infty}$ for $p = p_\theta$ was proved by Kruglov [Kr]).

We now turn to the notion of "k structure" which was mainly developed in the papers [DJL], [DL] and [F].

In what follows, we equip $I\!N^k$ or $\{1,2,\dots,n\}^k$ with the partial order inherited from the product structure, namely for i,j, in $I\!N^k$ we write $i < j$ if $i_1 \leq j_1$, $i_2 \leq j_2, \dots, i_k \leq j_k$.

Definition: A Banach space X has a global k-structure if there is a bounded collection $\{x_i \mid i \in I\!N^k\}$ in X and a bounded collection $\{x_i^* \mid i \in I\!N^k\}$ in X^* such that

$$\langle x_i^*, x_j\rangle = \begin{cases} 1 & \text{if } j < i \\ 0 & \text{otherwise}. \end{cases}$$

A space X has a local k-structure if there is a space Y with a global k-structure which is finitely representable in X (in the sense of [J3]). For the connections of this notion to reflexivity, we refer the reader to [DJL] and [DL]. We recall merely the following fact from [DJL]. Let $R_1(X) = X^{**}/X$, $R_2(X) = R_1(R_1(X)), \dots, R_k(X) = R_1(R_{k-1}(X))$, etc....

Then a space X has a local k-structure iff there is a space Y which is finitely representable into X and such that $R_k(X) \neq \{o\}$. For $k = 1$, this is essentially due to James cf. [J6],[J3],[Pt].

Consider $x = (x_i)$ in $I\!R^{I\!N^k}$.

For $i = (i_1, \dots, i_k)$ let

$$\Delta_i x = \sum_\omega x_{i_1-\omega_1, i_2-\omega_2, \dots, i_k-\omega_k} \cdot (-1)^{\omega_1+\dots+\omega_k}$$

where the sum runs over all ω in $\{0,1\}^k$. (Here we always make the convention that $x_i = 0$ if one of the indices $i_1, \dots, i_k$ is negative).

We define $v_1[k]$ as the space of those x such that $\sum_{i \in I\!N^k} |\Delta_i x| < \infty$, equipped with the norm $\|x\| = \sum_{i \in I\!N^k} |\Delta_i x|$.

We also denote by $j[k]$ the natural inclusion map

$$v_1[k] \subset \ell_\infty(\mathbb{N}^k) .$$

The reader can check easily that $v_1[k]$ can be identified with the projective tensor product $v_1 \hat{\otimes} v_1 \otimes \ldots \hat{\otimes} v_1$ (k times).

Then it is easy to check that a space X has a global k-structure iff the map $j[k]$ factors through X.

Our main result concerning k-structure is the existence for each $q > 2$ and each k of a space of type 2 and cotype q which has a global k-structure. This is easy to derive from the preceding study of the spaces $A_{\theta q}(B)$ with B a Banach space. Here is how to proceed: we may consider the case when $B = A_{\theta q}$, in that case we denote $A_{\theta q}[2] = A_{\theta q}(A_{\theta q})$.

More generally, let $A_{\theta q}[2](B) = A_{\theta q}(A_{\theta q}(B))$. Then, we define inductively

$$A_{\theta q}[k+1] = A_{\theta q}(A_{\theta q}[k])$$

and

$$A_{\theta q}[k+1](B) = A_{\theta q}(A_{\theta q}[k](B)) .$$

It is then easy to check that $v_1[k] \subset A_{\theta q}[k] \subset \ell_\infty(\mathbb{N}^k)$ so that all the spaces $A_{\theta q}[k]$ have a global k-structure.

Now using Remark 9 and Lemma 10, one can check (we leave the details to the reader) that if $\theta \neq \frac{1}{2}$ $A_{\theta q}[k]$ is of type $2 \wedge p_\theta \wedge q$ and cotype $2 \vee p_\theta \vee q$. This imples the result claimed above, which improves the work of Farahat [F]. Note that it follows from Farahat's work that there is for every k a space with a global k-structure and no $(k+1)$-structure (either local or global). We can also derive this fact in a somewhat more explicit manner. This follows by examining carefully the spaces $A_{\theta q}[k]$ for $\theta = \frac{1}{2}$, $q = 2$. In that case, we denote simply

$$A[k] = A_{\frac{1}{2}2}[k]$$

and

$$A[k](B) = A_{\frac{1}{2}2}[k](B) .$$

For a Banach space X, let

$$\gamma_2^n(X) = \sup\{d(E, \ell_2^n) \quad E \subset X \quad \dim E = n\} .$$

One can prove the following two lemmas (in a similar way as in the case $k = 1$).

Lemma 12: There is a constant C_k such that for all $n > 1$

$$\gamma_2^n(A[k]) \leq C_k(\mathrm{Log}\ n)^k \ .$$

Lemma 13: Let X be a Banach space. If $\gamma_2^n(X)$ is a $o((\mathrm{Log}\ n)^k)$, then the space X cannot have a local k-structure.

The proof of Lemma 13 follows by examining the γ_2-norm of $j[k]$ restricted to the space of the basis vectors with indices in $\{0, 1, \ldots, n\}^k$. This norm is (for k fixed) equivalent to $(\mathrm{Log}\ n)^k$ when $n \to \infty$.

It is then an immediate consequence of Lemmas 12 and 13 that $A[k]$ is a space with global k-structure and without any local $(k+1)$-structure.

We have thus obtained a generalization of Theorem 1(iii) for $k > 1$: There is a Banach space X such that $R_k(X) \neq \{o\}$ and such that $\gamma_2^n(X)$ if $O((\mathrm{Log}\ n)^k)$ when $n \to \infty$. This is sharp in the sense that if $\gamma_2^n(X)$ is $o((\mathrm{Log}\ n)^k)$ then (by the above lemmas) $R_k(X)$ must be $\{o\}$.

We can also give an application of our results to the theory of bases in Banach spaces, as follows.

Theorem 14: Any separable Banach space B of type $p > 1$ and of cotype $r > 2$ is isomorphic to a quotient of a space X which has a basis and is of type p and of cotype r.

Proof: We will consider $A_{\theta q}(B)$. Let $P_n : A_{\theta q}(B) \to A_{\theta q}(B)$ be the operator defined by

$$\forall x = (x_n) \qquad P_n x = (x_0, x_1, \ldots, x_n, x_n, x_n, \ldots)$$

(x_n is repeated after the n-th place).

By interpolation (since this holds for $v_1(B)$ and $\ell_\infty(B)$) we have $\sup_n \|P_n\| < \infty$.

Moreover, there is a natural operator

$$\Phi : c(B) \to B$$

defined by $\Phi(x) = \lim_{n \to \infty} x_n$.

Note that Φ realizes B "simultaneously" as a quotient of $v_1(B)$ and of $c(B)$.

Now let $\{y_n\}$ be a sequence in B which is dense in the unit ball of B.

Let $\overline{y}_n \in v_1(B)$ be defined by

$$\overline{y}_n = (0, 0, \ldots 0, y_n, y_n, y_n \ldots)$$

(the j-th coordinate is 0 if $j < n$ and y_n if $j \geq n$).

Note that $P_k \overline{y}_n = \overline{y}_n$ if $k \geq n$ and $P_k \overline{y}_n = 0$ if $k < n$. Since $\sup_n \|P_n\| < \infty$, this immediately implies that $\{\overline{y}_n\}$ is a basic sequence (cf. [LT1]) in $A_{\theta q}(B)$. Let X be the closed span of this sequence in $A_{\theta q}(B)$. Then X has clearly a basis and moreover since $\Phi(\overline{y}_n) = y_n$, $\Phi_{|X}$ is a quotient mapping.

Finally, by Corollary 7, we can choose $q = 2$ and $p < p_\theta < r$ so that X is of type p and of cotype r.

Remark: If B is of type $p < 2$ with $p > 1$ and of cotype 2, we can find X as in Theorem 14 but of type p and of cotype 2 (same proof as above with $p < p_\theta < 2$).

Remark: In some cases, one can perhaps get a result similar to Theorem 14 but concerning *embeddings* into a space with a basis. In the finite dimensional case we have the following statement (same proof as Theorem 14).

For $1 < p \leq 2 < r$, there is a constant C with the following property. Any finite dimensional Banach space B with $n = \dim B$ embeds isometrically into a space X with $\dim X \leq C^n$ which has a basis with basis constant $\leq C$ and which satisfies $T_p(X) \leq C T_p(B)$ and $C_r(X) \leq C C_r(B)$.

Remark: Actually, one can also derive a weak form of Theorem 14 from Farahat's work [F].

We now give some applications of our results to Gaussian random variables and Gaussian processes. Recall that a Gaussian process $\{X_t \mid t \in T\}$ is a collection of random variables such that all the linear combinations of $\{X_t \mid t \in T\}$ are Gaussian.

Let (g_n) be an i.i.d. sequence of standard normal Gaussian random variables. All the random variables which appear below are assumed to be defined on some probability space $(\Omega, \mathcal{A}, I\!P)$. By a known result (cf. [MP] Corollary 1.3, p. 68) or by repeating word for word the same argument, the above Theorem 8 remains valid with the sequence $\{g_n\}$ instead of the sequence $\{\varepsilon_n\}$. We thus derive easily the following formally more general statement.

Theorem 15: Let $\{X_n \mid n \in I\!N\}$ be a Gaussian process. Let H be the closed span in L_2 of the sequence $\{X_n \mid n \in I\!N\}$. Let $0 < \theta < 1$, $1 \leq q < \infty$. Then the sequence $\{X_n(\omega) \mid n \in I\!N\}$ belongs ω-almost surely to $A_{\theta q}$ iff the sequence $\{X_n \mid n \in I\!N\}$ belongs to $A_{\theta q}(H)$.

In the next statement, we will use the following notation. For a Banach space B and an element $x = (x_n)_{n \geq 0}$ in $B^{I\!N}$, let

$$\lambda_k(x) = \sup \frac{1}{k} \sum_{j=1}^{j=k} \|x_{n_j} - x_{n_{j-1}}\|$$

where the supremum runs over all increasing sequences $n_0 \leq n_1 \leq \ldots \leq n_k$ of k integers.

It was proved in [BP] that the K_t-functional for the couple $(v_1(B), \ell_\infty(B))$ of an element x in $v_1(B) + \ell_\infty(B)$ is uniformly equivalent for $t = k$ with $k\lambda_k(x)$. (The proof is similar to that of Lemma 2(ii) above).

Therefore, the quantity $\left(\int_0^\infty (t^{-\theta} K_t)^q \frac{dt}{t}\right)^{1/q}$, i.e., the norm of x in $A_{\theta q}$, is equivalent to the norm of $\{\lambda_k(x)\}$ in $\ell_{p_\theta q}$.

This result is sometimes useful to check that a concrete x belongs to $A_{\theta q}$.

Returning to the situation of Theorem 15 and using the inclusion $A_{\theta p} \subset v_p$ for $p = p_\theta$, we obtain immediately

Corollary 16: Let $1 < p < \infty$. Let $\{X_n\}$ be as above. We consider the property "$\{X_n(\omega) \mid n \in \mathbb{N}\}$ belongs ω-a.s. to v_p".

For this property to hold, it is sufficient that $\{X_n \mid n \in \mathbb{N}\}$ belongs to $A_{\theta p_\theta}(H)$ and it is necessary that it belongs to $v_p(H)$. More explicitly, let $X = \{X_n\} \in H^{\mathbb{N}}$. Then the sufficient condition is equivalent to

$$\{\lambda_k(X)\} \in \ell_{p_\theta} , \tag{19}$$

and the necessary condition implies

$$\{\lambda_k(X)\} \in \ell_{p_\theta \infty} . \tag{20}$$

The reader should note that conditions such as (19) or (20) are often easy to prove or disprove. As an illustration we state the following.

Corollary 17: Let $\{g_n\}$ be an i.i.d. Gaussian sequence as above, let $\{\alpha_n\}$ be a sequence of real coefficients and let $S = (S_n)_{n \geq 0}$ be the sequence of the partial sums

$$S_n = \sum_{j \leq n} \alpha_j g_j .$$

Let $0 < \theta < \frac{1}{2}$, $1 \leq q < \infty$. Then S belongs to $A_{\theta q}$ iff $(\alpha_n)_{n \geq 0}$ belongs to $\ell_{p_\theta q}$.

The "only if" part is easy. Let us prove the "if" part. By Theorem 15, it suffices to prove that S belongs to $A_{\theta q}(H)$ if $(\alpha_n)_{n \geq 0}$ is in $\ell_{p_\theta q}$.

Let T be the linear operator $(\alpha_n)_{n \geq 0} \to (S_n)_{n \geq 0}$, considered as an operator from ℓ_1 into $v_1(H)$.

Clearly T is bounded also from ℓ_2 into $v_2(H)$, hence, by interpolation, T is bounded from $(\ell_1, \ell_2)_{\eta q}$ into $(v_1(H), v_2(H))_{\eta q}$ for $\frac{1}{p_\theta} = \frac{1-\eta}{1} + \frac{\eta}{2}$.

By Lemma 2 and the reiteration theorem, $(v_1(H), v_2(H))_{\eta q}$ coincides with $A_{\theta q}(H)$, hence T is bounded from $\ell_{p_\theta q}$ into $A_{\theta q}(H)$.. This proves the "if" part. q.e.d.

Remark: The preceding result implies that $(\alpha_n)_{n\geq 0} \in \ell_p$ is a necessary and sufficient condition for $S = (S_n)_{n\geq 0}$ to be a.s. in v_p, for $1 < p < 2$. Essentially this is proved in [Br], but the above proof is much simpler than the one in [Br]. Actually, a similar proof was shown to the first author by B. Maurey around 1974. The reader will note that the above argument is valid for a large class of non Gaussian random variables also. Indeed, the fact that $(\alpha_n) \in \ell_p \Rightarrow (S_n) \overset{\text{a.s.}}{\in} v_p$ can also be deduced from the fact that $A_{\theta p}$ is of type p $(1 < p < 2,\ p = p_\theta)$.

Using this, one can show the following generalization: Let $\{Y_n\}$ be a sequence of independent symmetric real valued random variables. Let $S_n = \sum_{k\leq n} Y_k$ and let $\Phi = (\sum |Y_k|^p)^{1/p}$ and $\Psi = \|\{S_n\}\|_{v_p}$. (Note that $\Psi \geq \Phi$.) Assume that $1 < p < 2$. One can then prove (using similar arguments) that for all $0 < r < \infty$ we have $\Psi \in L_r$ iff $\Phi \in L_r$, and the quantities $\|\Psi\|_r$ and $\|\Phi\|_r$ are equivalent. If $r \geq 1$, one can assume (instead of the symmetry) that the Y_n's are all of mean zero. Moreover, the preceding statements can be generalized for sequences $\{Y_n\}$ of independent mean zero random variables with values in the Banach space B, as follows. Assume that B is of type p_0 and let as before $\Phi = (\sum \|Y_k\|^p)^{1/p}$ and $\Psi = \|\{S_n\}\|_{v_p(B)}$, then, if $1 < p < p_0$, the norms of Φ and Ψ in L_r $(0 < r < \infty)$ are equivalent.

As we have mentioned in the introduction, the preceding results can be developed also for the function spaces V_p relative to $[0,1]$ or to $\mathbb{R}$ instead of the sequence space v_p (relative to $\mathbb{N}$).

In that case, it is known that the (function) spaces $A_{\theta q}$ can be described in terms of approximation by splines (cf. [BP]). Actually, the same idea can be easily adapted to our earlier setting of sequence spaces. By a "spline with k knots" we will mean a sequence $x = (x_n)_{n\geq 0}$ in $\mathbb{R}^{\mathbb{N}}$ such that $\mathbb{N}$ can be decomposed as the union of k (possibly unbounded) intervals on each of which x is constant. Let S_k be the set of all such x. We then define for any x in ℓ_∞

$$S_k(x) = \inf\{\|x - y\|_\infty \mid y \in S_k\} .$$

One can then prove the following result (similar results are in [BP], see also [BL] Chap. 7).

Theorem 18: Assume $o < \theta < 1$, $1 \leq q \leq \infty$. Let $x \in \ell_\infty$. Then x belongs to $A_{\theta q}$ iff $(S_k(x))_{k\geq 1}$ belongs to $\ell_{p_\theta q}$.

Proof: By the same argument as in Lemma 2.ii one can show that $S_k(x) \le C\lambda_k(x)$ for some constant C. This shows that (cf. remark preceding Corollary 16)

$$\|\{S_k\}\|_{p_\theta q} \le C\|x\|_{A_{\theta q}}$$

for some constant C.

Let us sketch the proof of the converse. This type of argument is essentially known. We work with $k = 2^n$. For each n, there is x^n in ℓ_∞ such that

$$\|x - x^n\|_\infty \le 2S_{2^n}(x) \ .$$

Let $a_n = S_{2^n}(x)$. Let $\Delta_n = x^n - x^{n+1}$ and $x^0 = 0$. Then we have

$$x = \sum_{n \ge 0} \Delta_n \quad \text{with} \quad \Delta_n \in S_{2^n + 2^{n+1}} \subset S_{2^{n+2}}$$

and $\|\Delta_n\|_\infty \le 4a_n$.

This implies that $K_t(x) = K_t(v_1, \ell_\infty; x)$ satisfies

$$\begin{aligned} K_t(x) &\le \|\sum_{n \le k} \Delta_n\|_{v_1} + t\|\sum_{n>k} \Delta_n\|_\infty \\ &\le \sum_{n \le k} 2^{n+3}\|\Delta_n\|_\infty + t\sum_{n>k}\|\Delta_n\|_\infty \qquad (21) \\ &\le \sum_{n\le k} 2^{n+4} a_n + 4t\sum_{n>k} a_n \ . \end{aligned}$$

This holds in partiuclar for $t = 2^k$. It is then easy to check from this that if $\{S_k\} \in \ell_{p_\theta q}$ then $\sum_n 2^{n\frac{q}{p}} a_n^q < \infty$ (for $p = p_\theta$) and hence by (21) $\sum \left(2^{-n\theta} K_{2^n}(x)\right)^q < \infty$. The latter is equivalent to $\int_o^\infty \left(t^{-\theta} K_t(x)\right)^q \frac{dt}{t} < \infty$. q.e.d.

We now briefly discuss the case of function spaces. More information will be available in [X2]. For $1 \le p \le \infty$. The space $V_p[0,1]$ can be defined as the space of all functions f on $[0,1]$ such that

$$\sup \left(|f(t_0)|^p + \sum_i |f(t_i) - f(t_{i-1})|^p\right)^{1/p} < \infty \qquad (22)$$

where the supremum runs over all sequences $t_0 < t_1 < \cdots$ in $[0,1]$. We denote by $\|f\|_{V_p}$ the expression in (22). A similar definition makes sense for $I\!R$ yielding the space $V_p(I\!R)$. One can also define similarly the spaces $V_p([0,1]; B)$ and $V_p(I\!R; B)$ for B a Banach space. The preceding results can then be generalized without any significant difficulty to this setting. The spaces $A_{\theta q}[0,1]$ can be defined as $\left(V_1[0,1], \ell_\infty([0,1])\right)_{\theta q}$ and similarly for $I\!R$.

We will quote only the following sample result, an application of our methods to the theory of Gaussian random processes, which generalizes Theorem 15.

Theorem 19: Assume $0 < \theta < 1$, $1 \le q < \infty$. Let $X = \left(X(t)\right)_{t\in[0,1]}$ be a Gaussian random process on a probability space $(\Omega, \mathcal{A}, I\!P)$. Let H be a closed span of $\{X(t) \mid t \in [0,1]\}$ in $L_2(\Omega, \mathcal{A}, I\!P)$. Then X has its sample paths almost surely in $A_{\theta q}[0,1]$ iff the function $t \to X(t)$ belongs to $A_{\theta q}([0,1]; H)$. Moreover, for X to have paths almost surely in $V_p[0,1]$ it is sufficient that X belongs to $A_{\theta p}([0,1]; H)$ (with $p = p_\theta$) and it is necessary that X belongs to $V_p(H)$ and a fortiori to $A_{\theta\infty}([0,1]; H)$.

Note: Here we do not distinguish between the process X and a version of X and we assume H (hence X) separable.

Remark: We can define $\lambda_k(X)$ in analogy with the above.

$$\lambda_k(X) = k^{-1} \sup \sum_{k=1}^{k} \|X(t_i) - X(t_{i-1})\|_2$$

where the sup runs over all increasing subsets $t_0 < \cdots < t_k$. It is then easy to check as above that X belongs to $A_{\theta q}(H)$ iff $\{\lambda_k(X)\}$ belongs to $\ell_{p_\theta q}$.

This condition is usually easy to prove or disprove. For instance in the case of Brownian motion, one can check easily that this holds if $\theta > \frac{1}{2}$ and $1 \le q < \infty$, but not for $\theta = \frac{1}{2}$, since we have simply $\lambda_k(X) \approx k^{-1/2}$ in that case.

One can generalize Corollary 17 (together with the remarks following it) to random processes with independent increments over $[0,1]$. We thus recover the results of [Br]. The processes for which the strong p-th variation (i.e., the expression (22)) is finite almost surely have been also considered by Millar (cf. [M]) and by Blumenthal and Getoor (cf. the references in [M]) but only in the case of independent increments. We refer the reader to [X2] for further developments.

After this paper was completed, we proved (using Theorem 3.1 in [BP]) a natural generalization of the above Theorem 3 with L_p replaced by an arbitrary subspace of L_p and L_r replaced by an arbitrary quotient of L_r. In particular, it follows that all the spaces $A_{\theta q}$ $(0 < \theta < 1,\ 1 \le q < \infty)$ possess the Gordon-Lewis property, meaning that every 1-absolutely summing operator on $A_{\theta q}$ with arbitrary range factors through L_1.

Acknowledgment: We would like to thank B. Maurey for several discussions with the first author on the subject of this paper, going back to the period 73/74. These discussions, although somewhat incomplete, contained many observations which are used freely throughout this paper.

References

[BL] J. Bergh, J.Löfström. Interpolation spaces. An Introduction. Springer-Verlag. Berlin-Heidelberg-New York (1976).

[BP] J. Bergh, J. Peetre. On the spaces $V_p(0 < p \leq \infty)$. Bollettino della Unione Matematica Italiana. 10 (1974) 632-648.

[B] J. Bretagnolle. p-variation des fonctions aléatoires. Séminaire de Probabilités VI (1972) Springer Lecture Notes No. 258, p. 51-71.

[BB] P.L. Butzer, H. Behrens. Semi-groups of operators and approximation. Springer-Verlag. Berlin-Heidelberg-New York (1967).

[DJL] W.J. Davis, W.B. Johnson and J. Lindenstrauss. The ℓ_1^n problem and degrees of non reflexivity. Studia Math. 55 (1976), 123-139.

[DL] W.J. Davis, J. Lindenstrauss. The ℓ_1^n problem and degrees of non-reflexivity II. Studia Math. 58 (1976) 179-196.

[DFJP] W.J. Davis, T. Figiel, W. Johnson, A. Pełczyński. Factoring weakly compact operators. Journal of Funct. Anal. 17 (1974) 311-327.

[F] J. Farahat. On the problem of k structure. Israel J. Math. 28 (1977) 141-150.

[H] T. Holmsted. Interpolation of quasi-normed spaces. Math. Scand 26 (1970) 177-199.

[J1] R.C. James. A nonreflexive Banach space that is uniformly nonoctahedral. Israel J. Math. 18 (1974) 145-155.

[J2] R.C. James. Nonreflexive spaces of type 2. Israel J. Math. 30 (1978) 1-13.

[J3] R.C. James. Some self-dual properties of normed linear spaces. Annals of Math Studies. No. 69 (1972) 159-175.

[J4] R.C. James. Uniformly nonsquare Banach spaces. Annals of Math. 80 (1964) 542-550.

[J5] R.C. James. Banach spaces quasi-reflexive of order one. Studia Math. 60 (1977) 157-177.

[J6] R.C. James. Characterizations of reflexivity. Studia Math. 23 (1964) 205-216.

[J7] R.C. James. Reflexivity and the supremum of linear functionals. Annals of Math. 66 (1957) 159-169.

[JL] R.C. James, J. Lindenstrauss. The octohedral problem for Banach spaces. Proceedings of the Seminar on Random Series, Convex Sets, and Geometry of Banach Spaces. Aarhus University (Denmark) 1974 p. 100-120.

[JT] W. Johnson, L. Tzafrari. Some more Banach spaces which do not have local unconditional structure. Houston J. Math. 3 (1977) 55-60.

[Ka] M.I. Kadeč. The superreflexivity property of a Banach space in terms of the closeness of its finite dimensional subspaces to Euclidean spaces. Functional analysis and its Applications 12 (1978) 142-144.

[Kr] A. Kruglov. On some spaces of means. Vestnik Leningr. Univ. 1 (1972) 155-156.

[KS] A. Kruglov, M. Solomjak. Interpolation of operators in the spaces V_p. Vestnik Leningrad Univ. (1971) No. 13. Translated Vol 4 (1977) 209-216.

[K] S. Kwapień. Isomorphic characterization of inner-product spaces by orthogonal series with vector valued coefficients. Studia Math. 44 (1972) 583-595.

[LP] J. Lindenstrauss, A. Pełczyński. Absolutely summing operators in L_p-spaces and their applications. Studia Math. 29 (1968) 275-326.

[LT1] J. Lindenstrauss, L. Tzafriri. Classical Banach spaces I. Springer-Verlag. Berlin, Heidelberg, New York (1977).

[LT2] J. Lindenstrauss, L. Tzafriri. Classical Banach spaces II. Springer-Verlag. Berlin, Heidelberg, New York (1979).

[MP] B. Maurey, G. Pisier. Séries de variables aléatoires vectorielles indépendantes et propriétés géométriques des espaces de Banach. Studia Math. 58 (1976), 45-90.

[M] P. Millar. Path behavior of processes with stationary independent increments. Z. für Wahrshein. 17 (1971), 53-73.

[Mi] M. Milman. Complex interpolation and geometry of Banach spaces. Ann. Mat. Pura Appl. 136 (1984), 317-328.

[P1] G. Pisier. Martingales with values in uniformly convex spaces. Israel J. Math 20 (1975), 326-350.

[P2] G. Pisier. Holomorphic semi-groups and the geometry of Banach spaces. Annals of Math. 115 (1982), 375-392.

[Pt] V. Ptak. Biorthogonal systems and reflexivity of Banach spaces. Čekoslovak Math. Journal 9 (1959), 319-326.

[TJ] N. Tomczak-Jaegermann. Computing 2-summing norm with few vectors. Arkiv för Matematik 17 (1979), 273-277.

[X1] Q. Xu. In this volume.

[X2] Q. Xu. In preparation

COTYPE OF THE SPACES $(A_0, A_1)_{\theta 1}$

Quanhua XU

Wuhan University
and
Université Paris 6

In this note, we compute the cotype of the interpolation spaces $(A_0, A_1)_{\theta 1}$ for $0 < \theta < 1$ when one of the spaces A_0, A_1 is K-convex. An example of Pisier shows that there exists a couple (A_0, A_1) such that A_1 is of cotype 2 but none of the spaces $(A_0, A_1)_{\theta q}$ $(0 < \theta < 1, 1 \leq q \leq \infty)$ is of finite cotype, (cf. [1]). Therefore, the "cotype" does not pass to the interpolation spaces, contrary to the case of type (cf. [4]). In the case $0 < \theta < 1$ and $1 < q < \infty$, it is known that if A_0 or A_1 is K-convex then $(A_0, A_1)_{\theta q}$ is also K-convex and its cotype is easy to compute using the duality with type. This procedure is however not applicable to the case $q = 1$, because $(A_0, A_1)_{\theta 1}$ is not K-convex except for the trivial case (cf. [2]). We will use a direct and very simple method to calculate the cotype of $(A_0, A_1)_{\theta 1}$.

Our terminology and notation are standard and we use the K-method of Peetre for interpolation. We have

Theorem: Suppose that (A_0, A_1) is a couple of Banach spaces and $0 < \theta < 1$. If A_1 is K-convex and of cotype q_1, then $(A_0, A_1)_{\theta 1}$ is of cotype q_θ, where $\frac{1}{q_\theta} = \frac{1-\theta}{\infty} + \frac{\theta}{q_1}$.

This theorem is an immediate corollary of the following general lemma.

Lemma: Suppose (A_0, A_1) is an interpolation couple of Banach spaces and $0 < \theta < 1$, $0 < p, q < \infty$. Then there exists a constant $C = C(\theta, p, q)$ such that for all finite sequences (x_i) in $A_0 \cap A_1$, we have

$$
\begin{aligned}
C^{-1} \| \sum \varepsilon_i x_i \|_{L^q((A_0, A_1)_{\theta q})} &\leq \| \sum \varepsilon_i x_i \|_{(L^p(A_0), L^p(A_1))_{\theta q}} \\
&\leq C \| \sum \varepsilon_i x_i \|_{L^q((A_0, A_1)_{\theta q})} ,
\end{aligned} \tag{1}
$$

where (ε_i) is the Rademacher sequence.

Proof: We denote by $K_t(A_0, A_1)$ the space $A_0 + A_1$ equipped with the norm $K_t(\cdot)$. We first prove the following useful property on K_t:

(*) the K_t-norm on $L^p(A_0) + L^p(A_1)$ is uniformly equivalent for $t > 0$ to the norm on $L_p(K_t(A_0, A_1))$.

In fact, for every $f \in L^p(A_0) + L^p(A_1)$, we have

$$\|f\|_{K_t(L^p(A_0),L^p(A_1))} = \inf\{\|f_0\|_{L^p(A_0)} + t\|f_1\|_{L^p(A_1)} \mid f = f_0 + f_1, f_i \in L^p(A_i),\ i = 0,1\}$$

uniformly for $t > 0$

$$\sim \inf\{[\int (\|f_0\|_{A_0} + t\|f_1\|_{A_1})^p d\mu]^{1/p} | f = f_0 + f_1,\ f_i \in L^p(A_i),\ i = 0,1\}$$
$$= \|f\|_{L^p(K_t(A_0,A_1))} \, .$$

By the Kahane inequalities we have

$$C_{p,q}^{-1}\|\sum \varepsilon_i x_i\|_{L^q(K_t(A_0,A_1))} \le \|\sum \varepsilon_i x_i\|_{L^p(K_t(A_0,A_1))}$$
$$\le C_{p,q}\|\sum \varepsilon_i x_i\|_{L^q(K_t(A_0,A_1))}$$

for all finite sequences (x_i) in $A_0 + A_1$, where $C_{p,q}$ is a constant depending only on p and q. Using $(*)$ we deduce from the above inequalities by a suitable integration that

$$C^{-1}\|\sum \varepsilon_i x_i\|_{L^q(A_0,A_1)_{\theta,q})} \le \|\sum \varepsilon_i x_i\|_{(L^p(A_0),L^p(A_1))_{\theta,q}}$$
$$\le C\|\sum \varepsilon_i x_i\|_{L^q((A_0,A_1)_{\theta,q})} \, ,$$

where $C = C(\theta, p, q)$ depends only on θ, p and q. This is the desired result

Proof of the Theorem. We will show more generally that $(A_0, A_1)_{\theta q}$ is of cotype $\max(q, q_\theta)$, for $1 \le q < \infty$. Taking $p = 2$ in (1) we get

$$\text{(2)} \qquad \|\sum \varepsilon_i x_i\|_{(L^2(A_0),L^2(A_1))_{\theta,q}} \le C\|\sum \varepsilon_i x_i\|_{L^q((A_0,A_1)_{\theta,q})}$$

for all $(x_i) \subset A_0 \cap A_1$. Since A_1 is K-convex and of cotype q_1, by a known result (cf. [3]) we see that the following operator T is bounded:

$$T : L^2(A_1) \longrightarrow \ell^{q_1}(A_1)$$
$$f \longmapsto (\int f\varepsilon_i \, d\mu)_i \, .$$

Clearly, $T : L^2(A_0) \to \ell^\infty(A_0)$ is also bounded. We obtain, therefore, by interpolation

$$\text{(3)} \qquad T : \big(L^2(A_0), L^2(A_1)\big)_{\theta,q} \longrightarrow \big(\ell^\infty(A_0), \ell^{q_1}(A_1)\big)_{\theta,q} \, .$$

We can evidently suppose $q \le q_\theta$ (otherwise, we choose $q_1' > q_1$ such that $\frac{1}{q} = \frac{1-\theta}{\infty} + \frac{\theta}{q_1'}$). In this case, we have the bounded inclusion

$$\text{(4)} \qquad \big(\ell^\infty(A_0), \ell^{q_1}(A_1)\big)_{\theta,q} \hookrightarrow \ell^{q_\theta}\big((A_0, A_1)_{\theta,q}\big) \, .$$

Combining (2), (3) and (4) shows that $(A_0, A_1)_{\theta,q}$ is of cotype q_θ, which completes the proof of the theorem.

REFERENCES

[1] S.J. Dilworth. Complex convexity and the geometry of Banach Spaces. Math. Proc. Camb. Phil. Soc. 99 (1986), 495-506.

[2] M. Lévy. Thèse de 3e cycle, Université Paris 6, Février 1980.

[3] G. Pisier. Holomorphic semi-groups and the geometry of Banach Spaces. Annals of Maths. 115 (1982), 375-392.

[4] J. Lindenstrauss and L. Tzafriri. Classical Banach spaces II. Function spaces. Chapter 2.g. Springer-Verlag (1979).

LECTURE NOTES IN MATHEMATICS
Edited by A. Dold and B. Eckmann

Some general remarks on the publication of monographs and seminars

In what follows all references to monographs, are applicable also to multiauthorship volumes such as seminar notes.

1. Lecture Notes aim to report new developments - quickly, informally, and at a high level. Monograph manuscripts should be reasonably self-contained and rounded off. Thus they may, and often will, present not only results of the author but also related work by other people. Furthermore, the manuscripts should provide sufficient motivation, examples and applications. This clearly distinguishes Lecture Notes manuscripts from journal articles which normally are very concise. Articles intended for a journal but too long to be accepted by most journals, usually do not have this "lecture notes" character. For similar reasons it is unusual for Ph.D. theses to be accepted for the Lecture Notes series.

   Experience has shown that English language manuscripts achieve a much wider distribution.

2. Manuscripts or plans for Lecture Notes volumes should be submitted either to one of the series editors or to Springer-Verlag, Heidelberg. These proposals are then refereed. A final decision concerning publication can only be made on the basis of the complete manuscripts, but a preliminary decision can usually be based on partial information: a fairly detailed outline describing the planned contents of each chapter, and an indication of the estimated length, a bibliography, and one or two sample chapters - or a first draft of the manuscript. The editors will try to make the preliminary decision as definite as they can on the basis of the available information.

3. Lecture Notes are printed by photo-offset from typed copy delivered in camera-ready form by the authors. Springer-Verlag provides technical instructions for the preparation of manuscripts, and will also, on request, supply special staionery on which the prescribed typing area is outlined. Careful preparation of the manuscripts will help keep production time short and ensure satisfactory appearance of the finished book. Running titles are not required; if however they are considered necessary, they should be uniform in appearance. We generally advise authors not to start having their final manuscripts specially tpyed beforehand. For professionally typed manuscripts, prepared on the special stationery according to our instructions, Springer-Verlag will, if necessary, contribute towards the typing costs at a fixed rate.

   The actual production of a Lecture Notes volume takes 6-8 weeks.

.../...

4. Final manuscripts should contain at least 100 pages of mathematical text and should include

   - a table of contents
   - an informative introduction, perhaps with some historical remarks. It should be accessible to a reader not particularly familiar with the topic treated.
   - subject index; this is almost always genuinely helpful for the reader.

5. Authors receive a total of 50 free copies of their volume, but no royalties. They are entitled to purchase further copies of their book for their personal use at a discount of 33 1/3 %, other Springer mathematics books at a discount of 20 % directly from Springer-Verlag.

   Commitment to publish is made by letter of intent rather than by signing a formal contract. Springer-Verlag secures the copyright for each volume.

Vol. 1117: D.J. Aldous, J.A. Ibragimov, J. Jacod, Ecole d'Été de Probabilités de Saint-Flour XIII – 1983. Édité par P.L. Hennequin. IX, 409 pages. 1985.

Vol. 1118: Grossissements de filtrations: exemples et applications. Seminaire, 1982/83. Edité par Th. Jeulin et M. Yor. V, 315 pages. 1985.

Vol. 1119: Recent Mathematical Methods in Dynamic Programming. Proceedings, 1984. Edited by I. Capuzzo Dolcetta, W.H. Fleming and T. Zolezzi. VI, 202 pages. 1985.

Vol. 1120: K. Jarosz, Perturbations of Banach Algebras. V, 118 pages. 1985.

Vol. 1121: Singularities and Constructive Methods for Their Treatment. Proceedings, 1983. Edited by P. Grisvard, W. Wendland and J.R. Whiteman. IX, 346 pages. 1985.

Vol. 1122: Number Theory. Proceedings, 1984. Edited by K. Alladi. VII, 217 pages. 1985.

Vol. 1123: Séminaire de Probabilités XIX 1983/84. Proceedings. Edité par J. Azéma et M. Yor. IV, 504 pages. 1985.

Vol. 1124: Algebraic Geometry, Sitges (Barcelona) 1983. Proceedings. Edited by E. Casas-Alvero, G.E. Welters and S. Xambó-Descamps. XI, 416 pages. 1985.

Vol. 1125: Dynamical Systems and Bifurcations. Proceedings, 1984. Edited by B.L.J. Braaksma, H.W. Broer and F. Takens. V, 129 pages. 1985.

Vol. 1126: Algebraic and Geometric Topology. Proceedings, 1983. Edited by A. Ranicki, N. Levitt and F. Quinn. V, 423 pages. 1985.

Vol. 1127: Numerical Methods in Fluid Dynamics. Seminar. Edited by F. Brezzi, VII, 333 pages. 1985.

Vol. 1128: J. Elschner, Singular Ordinary Differential Operators and Pseudodifferential Equations. 200 pages. 1985.

Vol. 1129: Numerical Analysis, Lancaster 1984. Proceedings. Edited by P.R. Turner. XIV, 179 pages. 1985.

Vol. 1130: Methods in Mathematical Logic. Proceedings, 1983. Edited by C.A. Di Prisco. VII, 407 pages. 1985.

Vol. 1131: K. Sundaresan, S. Swaminathan, Geometry and Nonlinear Analysis in Banach Spaces. III, 116 pages. 1985.

Vol. 1132: Operator Algebras and their Connections with Topology and Ergodic Theory. Proceedings, 1983. Edited by H. Araki, C.C. Moore, Ş. Strătilă and C. Voiculescu. VI, 594 pages. 1985.

Vol. 1133: K.C. Kiwiel, Methods of Descent for Nondifferentiable Optimization. VI, 362 pages. 1985.

Vol. 1134: G.P. Galdi, S. Rionero, Weighted Energy Methods in Fluid Dynamics and Elasticity. VII, 126 pages. 1985.

Vol. 1135: Number Theory, New York 1983–84. Seminar. Edited by D.V. Chudnovsky, G.V. Chudnovsky, H. Cohn and M.B. Nathanson. V, 283 pages. 1985.

Vol. 1136: Quantum Probability and Applications II. Proceedings, 1984. Edited by L. Accardi and W. von Waldenfels. VI, 534 pages. 1985.

Vol. 1137: Xiao G., Surfaces fibrées en courbes de genre deux. IX, 103 pages. 1985.

Vol. 1138: A. Ocneanu, Actions of Discrete Amenable Groups on von Neumann Algebras. V, 115 pages. 1985.

Vol. 1139: Differential Geometric Methods in Mathematical Physics. Proceedings, 1983. Edited by H. D. Doebner and J. D. Hennig. VI, 337 pages. 1985.

Vol. 1140: S. Donkin, Rational Representations of Algebraic Groups. VII, 254 pages. 1985.

Vol. 1141: Recursion Theory Week. Proceedings, 1984. Edited by H.-D. Ebbinghaus, G.H. Müller and G.E. Sacks. IX, 418 pages. 1985.

Vol. 1142: Orders and their Applications. Proceedings, 1984. Edited by I. Reiner and K. W. Roggenkamp. X, 306 pages. 1985.

Vol. 1143: A. Krieg, Modular Forms on Half-Spaces of Quaternions. XIII, 203 pages. 1985.

Vol. 1144: Knot Theory and Manifolds. Proceedings, 1983. Edited by D. Rolfsen. V, 163 pages. 1985.

Vol. 1145: G. Winkler, Choquet Order and Simplices. VI, 143 pages. 1985.

Vol. 1146: Séminaire d'Algèbre Paul Dubreil et Marie-Paule Malliavin. Proceedings, 1983–1984. Edité par M.-P. Malliavin. IV, 420 pages. 1985.

Vol. 1147: M. Wschebor, Surfaces Aléatoires. VII, 111 pages. 1985.

Vol. 1148: Mark A. Kon, Probability Distributions in Quantum Statistical Mechanics. V, 121 pages. 1985.

Vol. 1149: Universal Algebra and Lattice Theory. Proceedings, 1984. Edited by S. D. Comer. VI, 282 pages. 1985.

Vol. 1150: B. Kawohl, Rearrangements and Convexity of Level Sets in PDE. V, 136 pages. 1985.

Vol 1151: Ordinary and Partial Differential Equations. Proceedings, 1984. Edited by B.D. Sleeman and R.J. Jarvis. XIV, 357 pages. 1985.

Vol. 1152: H. Widom, Asymptotic Expansions for Pseudodifferential Operators on Bounded Domains. V, 150 pages. 1985.

Vol. 1153: Probability in Banach Spaces V. Proceedings, 1984. Edited by A. Beck, R. Dudley, M. Hahn, J. Kuelbs and M. Marcus. VI, 457 pages. 1985.

Vol. 1154: D.S. Naidu, A.K. Rao, Singular Pertubation Analysis of Discrete Control Systems. IX, 195 pages. 1985.

Vol. 1155: Stability Problems for Stochastic Models. Proceedings, 1984. Edited by V.V. Kalashnikov and V.M. Zolotarev. VI, 447 pages. 1985.

Vol. 1156: Global Differential Geometry and Global Analysis 1984. Proceedings, 1984. Edited by D. Ferus, R.B. Gardner, S. Helgason and U. Simon. V, 339 pages. 1985.

Vol. 1157: H. Levine, Classifying Immersions into $\mathbb{R}^4$ over Stable Maps of 3-Manifolds into $\mathbb{R}^2$. V, 163 pages. 1985.

Vol. 1158: Stochastic Processes – Mathematics and Physics. Proceedings, 1984. Edited by S. Albeverio, Ph. Blanchard and L. Streit. VI, 230 pages. 1986.

Vol. 1159: Schrödinger Operators, Como 1984. Seminar. Edited by S. Graffi. VIII, 272 pages. 1986.

Vol. 1160: J.-C. van der Meer, The Hamiltonian Hopf Bifurcation. VI, 115 pages. 1985.

Vol. 1161: Harmonic Mappings and Minimal Immersions, Montecatini 1984. Seminar. Edited by E. Giusti. VII, 285 pages. 1985.

Vol. 1162: S.J.L. van Eijndhoven, J. de Graaf, Trajectory Spaces, Generalized Functions and Unbounded Operators. IV, 272 pages. 1985.

Vol. 1163: Iteration Theory and its Functional Equations. Proceedings, 1984. Edited by R. Liedl, L. Reich and Gy. Targonski. VIII, 231 pages. 1985.

Vol. 1164: M. Meschiari, J.H. Rawnsley, S. Salamon, Geometry Seminar "Luigi Bianchi" II – 1984. Edited by E. Vesentini. VI, 224 pages. 1985.

Vol. 1165: Seminar on Deformations. Proceedings, 1982/84. Edited by J. Ławrynowicz. IX, 331 pages. 1985.

Vol. 1166: Banach Spaces. Proceedings, 1984. Edited by N. Kalton and E. Saab. VI, 199 pages. 1985.

Vol. 1167: Geometry and Topology. Proceedings, 1983–84. Edited by J. Alexander and J. Harer. VI, 292 pages. 1985.

Vol. 1168: S.S. Agaian, Hadamard Matrices and their Applications. III, 227 pages. 1985.

Vol. 1169: W.A. Light, E.W. Cheney, Approximation Theory in Tensor Product Spaces. VII, 157 pages. 1985.

Vol. 1170: B.S. Thomson, Real Functions. VII, 229 pages. 1985.

Vol. 1171: Polynômes Orthogonaux et Applications. Proceedings, 1984. Edité par C. Brezinski, A. Draux, A.P. Magnus, P. Maroni et A. Ronveaux. XXXVII, 584 pages. 1985.

Vol. 1172: Algebraic Topology, Göttingen 1984. Proceedings. Edited by L. Smith. VI, 209 pages. 1985.

Vol. 1173: H. Delfs, M. Knebusch, Locally Semialgebraic Spaces. XVI, 329 pages. 1985.

Vol. 1174: Categories in Continuum Physics, Buffalo 1982. Seminar. Edited by F.W. Lawvere and S.H. Schanuel. V, 126 pages. 1986.

Vol. 1175: K. Mathiak, Valuations of Skew Fields and Projective Hjelmslev Spaces. VII, 116 pages. 1986.

Vol. 1176: R.R. Bruner, J.P. May, J.E. McClure, M. Steinberger, H_∞ Ring Spectra and their Applications. VII, 388 pages. 1986.

Vol. 1177: Representation Theory I. Finite Dimensional Algebras. Proceedings, 1984. Edited by V. Dlab, P. Gabriel and G. Michler. XV, 340 pages. 1986.

Vol. 1178: Representation Theory II. Groups and Orders. Proceedings, 1984. Edited by V. Dlab, P. Gabriel and G. Michler. XV, 370 pages. 1986.

Vol. 1179: Shi J.-Y. The Kazhdan-Lusztig Cells in Certain Affine Weyl Groups. X, 307 pages. 1986.

Vol. 1180: R. Carmona, H. Kesten, J.B. Walsh, École d'Été de Probabilités de Saint-Flour XIV – 1984. Édité par P.L. Hennequin. X, 438 pages. 1986.

Vol. 1181: Buildings and the Geometry of Diagrams, Como 1984. Seminar. Edited by L. Rosati. VII, 277 pages. 1986.

Vol. 1182: S. Shelah, Around Classification Theory of Models. VII, 279 pages. 1986.

Vol. 1183: Algebra, Algebraic Topology and their Interactions. Proceedings, 1983. Edited by J.-E. Roos. XI, 396 pages. 1986.

Vol. 1184: W. Arendt, A. Grabosch, G. Greiner, U. Groh, H.P. Lotz, U. Moustakas, R. Nagel, F. Neubrander, U. Schlotterbeck, One-parameter Semigroups of Positive Operators. Edited by R. Nagel. X, 460 pages. 1986.

Vol. 1185: Group Theory, Beijing 1984. Proceedings. Edited by Tuan H.F. V, 403 pages. 1986.

Vol. 1186: Lyapunov Exponents. Proceedings, 1984. Edited by L. Arnold and V. Wihstutz. VI, 374 pages. 1986.

Vol. 1187: Y. Diers, Categories of Boolean Sheaves of Simple Algebras. VI, 168 pages. 1986.

Vol. 1188: Fonctions de Plusieurs Variables Complexes V. Séminaire, 1979–85. Edité par François Norguet. VI, 306 pages. 1986.

Vol. 1189: J. Lukeš, J. Malý, L. Zajíček, Fine Topology Methods in Real Analysis and Potential Theory. X, 472 pages. 1986.

Vol. 1190: Optimization and Related Fields. Proceedings, 1984. Edited by R. Conti, E. De Giorgi and F. Giannessi. VIII, 419 pages. 1986.

Vol. 1191: A.R. Its, V.Yu. Novokshenov, The Isomonodromic Deformation Method in the Theory of Painlevé Equations. IV, 313 pages. 1986.

Vol. 1192: Equadiff 6. Proceedings, 1985. Edited by J. Vosmansky and M. Zlámal. XXIII, 404 pages. 1986.

Vol. 1193: Geometrical and Statistical Aspects of Probability in Banach Spaces. Proceedings, 1985. Edited by X. Femique, B. Heinkel, M.B. Marcus and P.A. Meyer. IV, 128 pages. 1986.

Vol. 1194: Complex Analysis and Algebraic Geometry. Proceedings, 1985. Edited by H. Grauert. VI, 235 pages. 1986.

Vol. 1195: J.M. Barbosa, A.G. Colares, Minimal Surfaces in $\mathbb{R}^3$. X, 124 pages. 1986.

Vol. 1196: E. Casas-Alvero, S. Xambó-Descamps, The Enumerative Theory of Conics after Halphen. IX, 130 pages. 1986.

Vol. 1197: Ring Theory. Proceedings, 1985. Edited by F.M.J. van Oystaeyen. V, 231 pages. 1986.

Vol. 1198: Séminaire d'Analyse, P. Lelong – P. Dolbeault – H. Skoda. Seminar 1983/84. X, 260 pages. 1986.

Vol. 1199: Analytic Theory of Continued Fractions II. Proceedings, 1985. Edited by W.J. Thron. VI, 299 pages. 1986.

Vol. 1200: V.D. Milman, G. Schechtman, Asymptotic Theory of Finite Dimensional Normed Spaces. With an Appendix by M. Gromov. VIII, 156 pages. 1986.

Vol. 1201: Curvature and Topology of Riemannian Manifolds. Proceedings, 1985. Edited by K. Shiohama, T. Sakai and T. Sunada. VII, 336 pages. 1986.

Vol. 1202: A. Dür, Möbius Functions, Incidence Algebras and Power Series Representations. XI, 134 pages. 1986.

Vol. 1203: Stochastic Processes and Their Applications. Proceedings, 1985. Edited by K. Itô and T. Hida. VI, 222 pages. 1986.

Vol. 1204: Séminaire de Probabilités XX, 1984/85. Proceedings. Edité par J. Azéma et M. Yor. V, 639 pages. 1986.

Vol. 1205: B.Z. Moroz, Analytic Arithmetic in Algebraic Number Fields. VII, 177 pages. 1986.

Vol. 1206: Probability and Analysis, Varenna (Como) 1985. Seminar. Edited by G. Letta and M. Pratelli. VIII, 280 pages. 1986.

Vol. 1207: P.H. Bérard, Spectral Geometry: Direct and Inverse Problems. With an Appendix by G. Besson. XIII, 272 pages. 1986.

Vol. 1208: S. Kaijser, J.W. Pelletier, Interpolation Functors and Duality. IV, 167 pages. 1986.

Vol. 1209: Differential Geometry, Peñíscola 1985. Proceedings. Edited by A.M. Naveira, A. Ferrández and F. Mascaró. VIII, 306 pages. 1986.

Vol. 1210: Probability Measures on Groups VIII. Proceedings, 1985. Edited by H. Heyer. X, 386 pages. 1986.

Vol. 1211: M.B. Sevryuk, Reversible Systems. V, 319 pages. 1986.

Vol. 1212: Stochastic Spatial Processes. Proceedings, 1984. Edited by P. Tautu. VIII, 311 pages. 1986.

Vol. 1213: L.G. Lewis, Jr., J.P. May, M. Steinberger, Equivariant Stable Homotopy Theory. IX, 538 pages. 1986.

Vol. 1214: Global Analysis – Studies and Applications II. Edited by Yu.G. Borisovich and Yu.E. Gliklikh. V, 275 pages. 1986.

Vol. 1215: Lectures in Probability and Statistics. Edited by G. del Pino and R. Rebolledo. V, 491 pages. 1986.

Vol. 1216: J. Kogan, Bifurcation of Extremals in Optimal Control. VIII, 106 pages. 1986.

Vol. 1217: Transformation Groups. Proceedings, 1985. Edited by S. Jackowski and K. Pawalowski. X, 396 pages. 1986.

Vol. 1218: Schrödinger Operators, Aarhus 1985. Seminar. Edited by E. Balslev. V, 222 pages. 1986.

Vol. 1219: R. Weissauer, Stabile Modulformen und Eisensteinreihen. III, 147 Seiten. 1986.

Vol. 1220: Séminaire d'Algèbre Paul Dubreil et Marie-Paule Malliavin. Proceedings, 1985. Edité par M.-P. Malliavin. IV, 200 pages. 1986.

Vol. 1221: Probability and Banach Spaces. Proceedings, 1985. Edited by J. Bastero and M. San Miguel. XI, 222 pages. 1986.

Vol. 1222: A. Katok, J.-M. Strelcyn, with the collaboration of F. Ledrappier and F. Przytycki, Invariant Manifolds, Entropy and Billiards; Smooth Maps with Singularities. VIII, 283 pages. 1986.

Vol. 1223: Differential Equations in Banach Spaces. Proceedings, 1985. Edited by A. Favini and E. Obrecht. VIII, 299 pages. 1986.

Vol. 1224: Nonlinear Diffusion Problems, Montecatini Terme 1985. Seminar. Edited by A. Fasano and M. Primicerio. VIII, 188 pages. 1986.

Vol. 1225: Inverse Problems, Montecatini Terme 1986. Seminar. Edited by G. Talenti. VIII, 204 pages. 1986.

Vol. 1226: A. Buium, Differential Function Fields and Moduli of Algebraic Varieties. IX, 146 pages. 1986.

Vol. 1227: H. Helson, The Spectral Theorem. VI, 104 pages. 1986.

Vol. 1228: Multigrid Methods II. Proceedings, 1985. Edited by W. Hackbusch and U. Trottenberg. VI, 336 pages. 1986.

Vol. 1229: O. Bratteli, Derivations, Dissipations and Group Actions on C*-algebras. IV, 277 pages. 1986.

Vol. 1230: Numerical Analysis. Proceedings, 1984. Edited by J.-P. Hennart. X, 234 pages. 1986.

Vol. 1231: E.-U. Gekeler, Drinfeld Modular Curves. XIV, 107 pages. 1986.

Vol. 1232: P.C. Schuur, Asymptotic Analysis of Soliton Problems. VIII, 180 pages. 1986.

Vol. 1233: Stability Problems for Stochastic Models. Proceedings, 1985. Edited by V.V. Kalashnikov, B. Penkov and V.M. Zolotarev. VI, 223 pages. 1986.

Vol. 1234: Combinatoire énumérative. Proceedings, 1985. Edité par G. Labelle et P. Leroux. XIV, 387 pages. 1986.

Vol. 1235: Séminaire de Théorie du Potentiel, Paris, No. 8. Directeurs: M. Brelot, G. Choquet et J. Deny. Rédacteurs: F. Hirsch et G. Mokobodzki. III, 209 pages. 1987.

Vol. 1236: Stochastic Partial Differential Equations and Applications. Proceedings, 1985. Edited by G. Da Prato and L. Tubaro. V, 257 pages. 1987.

Vol. 1237: Rational Approximation and its Applications in Mathematics and Physics. Proceedings, 1985. Edited by J. Gilewicz, M. Pindor and W. Siemaszko. XII, 350 pages. 1987.

Vol. 1238: M. Holz, K.-P. Podewski and K. Steffens, Injective Choice Functions. VI, 183 pages. 1987.

Vol. 1239: P. Vojta, Diophantine Approximations and Value Distribution Theory. X, 132 pages. 1987.

Vol. 1240: Number Theory, New York 1984–85. Seminar. Edited by D.V. Chudnovsky, G.V. Chudnovsky, H. Cohn and M.B. Nathanson. V, 324 pages. 1987.

Vol. 1241: L. Gårding, Singularities in Linear Wave Propagation. III, 125 pages. 1987.

Vol. 1242: Functional Analysis II with contributions by J. Hoffmann-Jørgensen et al. Edited by S. Kurepa, H. Kraljević and D. Butković. VII, 432 pages. 1987.

Vol. 1243: Non Commutative Harmonic Analysis and Lie Groups. Proceedings, 1985. Edited by J. Carmona, P. Delorme and M. Vergne. V, 309 pages. 1987.

Vol. 1244: W. Müller, Manifolds with Cusps of Rank One. XI, 158 pages. 1987.

Vol. 1245: S. Rallis, L-Functions and the Oscillator Representation. XVI, 239 pages. 1987.

Vol. 1246: Hodge Theory. Proceedings, 1985. Edited by E. Cattani, F. Guillén, A. Kaplan and F. Puerta. VII, 175 pages. 1987.

Vol. 1247: Séminaire de Probabilités XXI. Proceedings. Edité par J. Azéma, P.A. Meyer et M. Yor. IV, 579 pages. 1987.

Vol. 1248: Nonlinear Semigroups, Partial Differential Equations and Attractors. Proceedings, 1985. Edited by T.L. Gill and W.W. Zachary. IX, 185 pages. 1987.

Vol. 1249: I. van den Berg, Nonstandard Asymptotic Analysis. IX, 187 pages. 1987.

Vol. 1250: Stochastic Processes – Mathematics and Physics II. Proceedings 1985. Edited by S. Albeverio, Ph. Blanchard and L. Streit. VI, 359 pages. 1987.

Vol. 1251: Differential Geometric Methods in Mathematical Physics. Proceedings, 1985. Edited by P.L. García and A. Pérez-Rendón. VII, 300 pages. 1987.

Vol. 1252: T. Kaise, Représentations de Weil et GL_2 Algèbres de division et GL_n. VII, 203 pages. 1987.

Vol. 1253: J. Fischer, An Approach to the Selberg Trace Formula via the Selberg Zeta-Function. III, 184 pages. 1987.

Vol. 1254: S. Gelbart, I. Piatetski-Shapiro, S. Rallis. Explicit Constructions of Automorphic L-Functions. VI, 152 pages. 1987.

Vol. 1255: Differential Geometry and Differential Equations. Proceedings, 1985. Edited by C. Gu, M. Berger and R.L. Bryant. XII, 243 pages. 1987.

Vol. 1256: Pseudo-Differential Operators. Proceedings, 1986. Edited by H.O. Cordes, B. Gramsch and H. Widom. X, 479 pages. 1987.

Vol. 1257: X. Wang, On the C*-Algebras of Foliations in the Plane. V, 165 pages. 1987.

Vol. 1258: J. Weidmann, Spectral Theory of Ordinary Differential Operators. VI, 303 pages. 1987.

Vol. 1259: F. Cano Torres, Desingularization Strategies for Three-Dimensional Vector Fields. IX, 189 pages. 1987.

Vol. 1260: N.H. Pavel, Nonlinear Evolution Operators and Semigroups. VI, 285 pages. 1987.

Vol. 1261: H. Abels, Finite Presentability of S-Arithmetic Groups. Compact Presentability of Solvable Groups. VI, 178 pages. 1987.

Vol. 1262: E. Hlawka (Hrsg.), Zahlentheoretische Analysis II. Seminar, 1984–86. V, 158 Seiten. 1987.

Vol. 1263: V.L. Hansen (Ed.), Differential Geometry. Proceedings, 1985. XI, 288 pages. 1987.

Vol. 1264: Wu Wen-tsün, Rational Homotopy Type. VIII, 219 pages. 1987.

Vol. 1265: W. Van Assche, Asymptotics for Orthogonal Polynomials. VI, 201 pages. 1987.

Vol. 1266: F. Ghione, C. Peskine, E. Sernesi (Eds.), Space Curves. Proceedings, 1985. VI, 272 pages. 1987.

Vol. 1267: J. Lindenstrauss, V.D. Milman (Eds.), Geometrical Aspects of Functional Analysis. Seminar. VII, 212 pages. 1987.

This series reports new developments in mathematical research and teaching – quickly, informally and at a high level. The type of material considered for publication includes.

1. Research monographs
2. Lectures on a new field or presentations of a new angle in a classical field
3. Seminar work-outs
4. Reports of meetings, provided they are
   a) of exceptional interest and
   b) devoted to a single topic.

Texts which are out of print but still in demand may also be considered if they fall within these categories.

The timeliness of a manuscript is more important than its form, which may be unfinished or tentative. Thus, in some instances, proofs may be merely outlined and results presented which have been or will later be published elsewhere. If possible, a subject index should be included. Publication of Lecture Notes is intended as a service to the international mathematical community, in that a commercial publisher, Springer-Verlag, can offer a wide distribution of documents which would otherwise have a restricted readership. Once published and copyrighted, they can be documented in the scientific literature.

Manuscripts

Manuscripts should be no less than 100 and preferably no more than 500 pages in length.
They are reproduced by a photographic process and therefore must be typed with extreme care. Symbols not on the typewriter should be inserted by hand in indelible black ink. Corrections to the typescript should be made by pasting in the new text or painting out errors with white correction fluid. The typescript is reduced slightly in size during reproduction; best results will not be obtained unless on each page a typing area of 18 x 26.5 cm (7 x 10½ inches) is respected. On request, the publisher can supply paper with the typing area outlined. More detailed typing instructions are also available on request.

Manuscripts generated by a word-processor or computerized typesetting are in principle acceptable. However if the quality of this output differs significantly from that of a standard typewriter, then authors should contact Springer-Verlag at an early stage.

Authors of monographs and editors of proceedings receive 50 free copies.

Manuscripts should be sent to Prof. A. Dold, Mathematisches Institut der Universität Heidelberg, Im Neuenheimer Feld 288, 6900 Heidelberg, Germany; Prof. B. Eckmann, Eidgenössische Technische Hochschule, CH-8092 Zürich, Switzerland; or directly to Springer-Verlag Heidelberg.

Springer-Verlag, Heidelberger Platz 3, D-1000 Berlin 33
Springer-Verlag, Tiergartenstraße 17, D-6900 Heidelberg 1
Springer-Verlag, 175 Fifth Avenue, New York, NY 10010/USA
Springer-Verlag, 37-3, Hongo 3-chome, Bunkyo-ku, Tokyo 113, Japan

ISBN 3-540-18103-2
ISBN 0-387-18103-2